Bad Singer

Bad Singer

The Surprising Science of Tone Deafness and How We Hear Music

Tim Falconer

ANANSI

Published in Canada in 2016 by House of Anansi Press Inc.
www.houseofanansi.com

House of Anansi Press is committed to protecting our natural
environment. As part of our efforts, the interior of this book is printed on
paper that contains 100% post-consumer recycled fibres, is acid-free, and is
processed chlorine-free.

20 19 18 17 16 1 2 3 4 5

Library and Archives Canada Cataloguing in Publication

Falconer, Tim, 1958–, author
Bad singer : the surprising science of tone deafness and how we hear
music / Tim Falconer.

Includes bibliographical references and index.
Issued in print and electronic formats.
ISBN 978-1-77089-445-7 (bound).—ISBN 978-1-77089-446-4 (html)

1. Music—Acoustics and physics. 2. Musical perception. 3. Hearing.
4. Amusia. I. Title.

ML3820.F34 2016   781.1'1   C2015-907615-3
                         C2015-907616-1

Cover and text design: Alysia Shewchuk

Every reasonable effort has been made to trace ownership of copyright
material. The publisher will gladly rectify any inadvertent errors or
omissions in credits in future editions.

*We acknowledge for their financial support of our publishing program
the Canada Council for the Arts, the Ontario Arts Council, and the Government of
Canada through the Canada Book Fund.*

Printed and bound in Canada

For Carmen

CONTENTS

"We don't make music — it makes *us*."
—David Byrne, *How Music Works*

Music
and
Human
Evolution

"Sweet Dreams (Are Made of This)"

"So could you ask him to move or to stop singing?"

She was talking about me. Instead of sitting at our desks, we were standing along the side wall facing the middle of the classroom. We were in rows rather than in a clump as the teacher had arranged us, perhaps by height, perhaps by voice, perhaps by some other measure. The other details are even hazier. I don't remember the song or even why we were singing it. Maybe our teacher was trying to create a class choir for a school event. I was keen until the girl standing next to me complained that I was singing off-key and that it was throwing her off. *So, Mrs. Lennox, would you please do something about that?*

The memory ends abruptly there, as if I'd woken up from a bad dream before it was over, so I don't know what happened next. I think I'd remember if I felt particularly traumatized or went home in tears, though if I can still hear those words all these decades later, I guess it must have made some impression. But the experience was probably more confusing than anything else since I didn't understand what it meant to sing off-key. And I certainly didn't know how to go about singing on-key. Mostly, though, I was surprised. I was eight or nine and had no idea I was a bad singer. I don't remember singing in class again. Maybe Mrs. Lennox found something else for me to do. Or maybe I just went outside and played with my friends, which would have appealed to me more, anyway.

This is the part of the story where many people telling a similar tale would say, "And I never sang again." For a time, I thought that was true for me as well. But the more I mined my memory, the more I realized I was surprisingly slow to give up singing. I was a loudmouth kid, and I don't think I became all that self-conscious about my singing until I was about fifteen and on a ski trip to Vermont with some friends. Sitting around in the hotel, I absentmindedly started singing "Blackbird" by The Beatles. One of my friends—and I'm sure it's significant that she was a she—laughed and said, "Oh, Tim hit a note." And then everyone laughed. That's when I finally realized how bad I truly was.

After that, I never felt comfortable singing in public again. By which I mean singing in front of women or people I didn't know. The all-boys school I attended had mandatory morning prayers, which involved singing dusty old hymns. My buddies and I would scream out the lyrics at the top of our lungs. But that bad singing was an admirable act of rebellion as much as an expression of how much I enjoyed doing it. Today, I happily sing only when no one is around. Alone in the car, I'll belt out tune after tune. When I'm within earshot of other people, though, I am a silent crooner: an interior virtuoso, with or without iPod accompaniment, while walking down the street or riding my bike.

But I want to sing out loud—not like an angel, necessarily, but well enough that I'm not ashamed. When I'm at a friend's cottage and everyone brings out guitars around the campfire, I don't want to be the one who mumbles along quietly. And when the guys I play hockey with on Friday afternoons have a jam party with guitars and banjos, keyboards, drums, flutes, and harmonicas, I long to step up to the mic and croon away. I want to be the lead singer.

I'm a bad singer. And deep down, it matters. I've lived with the indignity and the frustration and the missed joy and assumed there was nothing I could do about it. But singing is something we should all be able to do, even if we do it badly. It is the most common form of music across all cultures, and traditionally, whenever people gathered they would sing. The voice is that rare instrument that we always have with us, so it's easy to create music whenever and wherever we want. And compared with our bare hands and feet—which make for serviceable percussion instruments—the voice is so much more versatile. We can sing a stirring aria, an angry punk song, or a tender lullaby.

Still, so many of us prefer to be silent. Our lack of confidence is understandable since we are at our most vulnerable when we sing. A poor guitarist still has the instrument as a shield; a singer has nothing to hide behind.

And yet, singing is what I most want to do.

In 2007, I spent a month at the Banff Centre, along with other artists, writers, and musicians working on their latest projects. My studio in the woods featured a black Blüthner baby grand piano, which I had to water. Seriously. That part of Alberta has an arid climate — it was an especially hot, dry summer, too — and pianos require a certain level of humidity to prevent the wood from warping. A couple of times a week, I filled a small plastic watering can in the kitchenette that mostly served as my wet bar. Then I stuck the can's long spout into the baby grand's internal humidifier and poured. I did this chore cheerfully, but otherwise, I never touched that beautiful instrument.

One day, as I was writing at my desk beside the piano, listening to music, as I always do when I work, I wondered how someone with such a large library of songs — ten thousand and growing rapidly — could be, as I had long suspected, tone deaf. I found that baffling. When I hosted a musicale at the studio, which was my way of ensuring the baby grand's keys didn't go untickled the whole time, several opera singers showed up. How could I be tone deaf and still love music? I asked. They scoffed at my self-diagnosis and one even offered to prove me wrong.

The next day, mezzo-soprano Catharin Carew showed me some breathing exercises, then asked me to match her notes. Her verdict: I wasn't singing the right ones, but I had a good, resonant voice. She didn't think I was tone deaf. I could differentiate notes, and when I sang the wrong one I was off by what she considered a perfect amount. "Instead of being ninety-seven cents off," she said, "you're exactly a dollar off."

I had no sense of why that was a good thing, but Carew seemed pleased, so I eagerly repeated her explanation to anyone who would listen, the way kids share jokes they don't really understand. Best of all, Carew told me that, with training, I had the potential to play the lead in a community theatre musical. (Later, in the dining hall, though, when I sat with the opera singers, she told them that I could be *in* community theatre.) I didn't care about a theatre career, and I had long ago come to accept that I'd never be a rock star. But I did want to be able to sing in front of other people without being a horrific, or hilarious, spectacle.

For a couple of years, the idea of learning to sing remained just one of the many things I dreamed of but never did anything about. Then a friend suggested I talk to Micah Barnes. The tall, well-built singing coach had a mop of dark hair and a small soul patch when I first met him. He owns a lot of vinyl; among the albums he had off the shelf and on display that day was *The Supremes Sing Holland-Dozier-Holland*. I really just wanted to hang out and listen to that record, but I forced myself to explain why I was there. I told him that I might be tone deaf, but I still wanted to learn to sing.

Barnes sat down on his piano bench and hummed a note, then asked me to try to match it. He didn't wince. We did it again. He pointed out that I was finding the right note, though not right away, so we'd need to work on speeding up that mental process. He told me I could sing harmony, though I didn't have the melody yet. When he asked me if I realized I was singing in harmony — this is what Carew's dollar metaphor was all about — I admitted I never have any idea what notes I'm singing. Singing for me, I explained, is like the time I went skeet shooting with my grandfather when I was a little kid. He gave me a shotgun and barked, "Pull." After a couple of my hopeless efforts, he took the gun away and said, "You're just shooting blindly."

I sing blindly, too.

Barnes understood my story. But he still asked me to sing a song.

"Oh, I can't sing," I insisted.

"Okay," he said. "That's a good place to start."

Once again, he asked me to sing a song. But my mind went completely blank. All the lyrics that usually come so easily disappeared. I'd told him that my dream was to be able to sing a Nick Lowe song on stage, but suddenly I couldn't remember more than the first line of "Rome Wasn't Built in a Day." So he sang "Amazing Grace" and I repeated his lines, even trying to ape the little flourish he put on "wretch."

He didn't think I had any "tone impairment," and suggested I come in once a week, maybe even twice, and then after a month, we could develop a plan to get me singing in public in a year, maybe less. I'd definitely need to work hard, he said, but he thought I could learn to sing. That was something I had never thought possible. And at times over the next several years, it didn't seem possible.

As it turned out, though, trying to learn how to sing proved to be a far more valuable experience than I ever could have imagined.

"It's Only Rock 'n' Roll (But I Like It)"

I felt ridiculous. I was only a few minutes into my first session with Micah Barnes, and I was already giggling nervously. He'd made me lie down on my back on his couch and make weird noises. After asking me to sing "Amazing Grace" a couple of times, he was hassling me about my breathing, trying to get me to do it with my diaphragm instead of my chest.

"Fill your diaphragm with air to make a balloon," he said, "and press your fingers against the balloon." But when I took in a breath my shoulders rose, which meant that I wasn't doing it right.

So he asked me to lie down and sigh repeatedly. "Put your hand on your diaphragm and as you breathe in, work against gravity and then let go. Feel how that's like a balloon that fills up and lets go? That's the secret."

"But that's not helping my ear," I said.

"You don't know that yet. Blow up a balloon and then let it go so it goes *aaah*. Oh, and do it with your tongue out."

Barnes, who was born in 1960, rents an apartment above a men's clothing store on a lively Toronto strip of bars, restaurants, and galleries called West Queen West. I'd expect a much younger man to live here. Streetcars rumble by regularly; occasionally fire trucks from the nearby station wail their sirens. When I started my lessons in 2011, he had a large black sectional couch from Ikea, a curved white chair from Structube, and a black bookcase full of CDs, LPs, photos, and keepsakes. He's not shy about showing off his favourites: he often displays albums and, for a long time, a copy of a songwriting magazine with Elvis Costello on the cover rested on a music stand. An upright piano, borrowed from a friend, sat by a window, overlooking the street. The place had no air conditioning, so in the summer he set up a fan and left the windows open, which made me especially self-conscious about appalling (or amusing) everyone on the sidewalk below. I'd been here once before, a year and a half earlier, when I'd first contacted him about my nutty idea to try to learn how to sing. Now, at long last, I was back, chequebook in hand and ready to do it.

Unlike me, Barnes comes from a musical tradition. His father, Milton Barnes, was a classical composer, and his younger brothers also went into the family business. Daniel is a drummer and bandleader, and Ariel is a cellist with the Vancouver Symphony Orchestra. In the 1980s, Micah played gigs as part of the Micah Barnes Trio, a cabaret act that included his brother Daniel, while also working in film, TV, and theatre. Realizing that the actors who were supposed to deliver his lyrics weren't good enough singers, he started inviting them to his place so he could offer a little informal training—and soon discovered that he had a knack for helping people free their minds and bodies so they could become better singers. And like many artists, he also figured out that teaching is a good way to supplement an income that can be meagre or sporadic or both.

In 1989, he joined The Nylons. A Canadian institution, this popular a cappella group has survived many lineup changes since 1979. Hired to sing baritone, Barnes soon understood that the job also meant singing tenor, including falsetto, and bass, depending on which of the four Nylons was singing lead. He was too busy to do much teaching, but he learned how crucial good technique is. Barnes lasted with the group until 1994, when the touring became too much of a drag. He moved to LA, where he kept playing and teaching. In 2003, he and house music producers Thunderpuss released a single called "Welcome to My Head" that hit number one on Billboard's club chart.

He's operated his one-teacher school, called Singers Playground, since 1996. Most of his clients are professional singers and actors (I was probably more impressed than I should have been when I learned that he'd worked with *Orphan Black*'s Tatiana Maslany) or people who want to be pros. For most amateurs—the folks who say, I've always wanted to sing and I'm finally going to take lessons—he's too expensive, charging $125 per session (or $400 for four).

After doing the breathing exercises for a while, I asked, "What do your neighbours think?"

"They know what I do for a living," he responded

matter-of-factly. And he makes even good singers do this exercise, because it gets them relaxed. If you're relaxed, he assured me, you'll hit the notes better. Maybe not perfectly, but better. That's how babies breathe — they use their stomach, not their chest. But as we get older, we unlearn the correct way and learn the wrong way: tension-filled breathing.

I kept doing as he asked, but I couldn't get over the giggles. Part of it was nervous tittering, but the whole situation was so ridiculous that I couldn't help but laugh at myself. He wondered what was up. "I looked at you and started laughing," I said. "I won't do that again."

While Barnes grew up in a musical home, I didn't. My mother didn't even sing to me when I was a baby. But she did play DJ for my sisters and me when we were little, even spinning records for us on an all-request basis. As a small boy, my favourite song was Peter, Paul and Mary's "Puff (the Magic Dragon)," which I insisted she play over and over again so I could sing along.

My father's vinyl collection was limited: bagpipe records; Christmas carols, which he played even in the summer; and *Whipped Cream & Other Delights* by Herb Alpert's Tijuana Brass. (I don't remember him ever playing that record. He may have owned it only for the seriously great cover featuring a beautiful woman slathered in whipped cream.)

My parents did make an effort. They signed my sisters up for the obligatory middle-class piano lessons. I started drumming classes but then had mysterious health problems — diagnosed several years later as childhood migraines — and never returned. My second-youngest sister took guitar lessons, but when the teacher tried to show her how to tune it, he was astonished that she couldn't hear the difference between pitches. She couldn't distinguish the frequency of notes (how high or low they are). Later, when my youngest sister took up guitar as an adult for a few years, she too was frustrated by her inability to hear pitch well enough to tune the instrument and eventually gave up trying. There wasn't a musician in the bunch.

So we did not make music together as a family. My mom figured we were all tone deaf, so what would be the point? And I don't remember my friends or cousins making a lot of music with their families. I'm sure some did, if only sing-alongs on road trips, when I wasn't around. But it was the 1960s and my generation—Generation Jones—grew up around the television, the electronic babysitter, not around the piano.

Still, I loved music. At twelve, I went to my first concert, an all-day, all-Canadian affair at Toronto's Varsity Stadium headlined by The Guess Who. By sixteen, I wanted to be Marvin Gaye. In the late '70s, I pogoed at punk shows. I was not the coolest kid—I was the only person in a button-down Oxford shirt and chinos in the mosh pit at a Viletones gig at Montreal's Hotel Nelson, and I'm still not sure if that punked-out woman who seemed so fascinated by my presence was hitting on me or goofing on me—but I loved giving myself over to what I was hearing. Still do.

Music means the most to us when we are in our teens and early twenties. That's when we have the time and the angsty need to connect with our emotions and with other people experiencing them. One U.S. study found that fifteen- to eighteen-year-olds spend just over three hours a day listening to tunes, with girls devoting more time to the activity than boys, and black and Hispanic kids more dedicated than white ones. For many people, music loses some of its power as they grow up. They don't have the time to seek out new artists — when I first began to notice this tendency among my friends, I joked about people whose last new album was Michael Jackson's *Thriller* — and seem content to buy overpriced tickets to see tired reunion shows in hockey arenas and football stadiums.

Regardless of our age, the way we listen to music has changed. As a teenager, I'd take some money I'd saved or earned and go downtown to the flagship Sam the Record Man store on Yonge Street or, later, while going to school in Montreal, I'd head over to Phantasmagoria on Park Avenue. I'd browse for a long time, flipping through the albums, buy one (ideally, more than one, if I was flush), and then go home to play it. I wouldn't just play it, I'd listen to it with focus and enthusiasm. I didn't try to multi-task other than studying the liner notes; the music fully captured my attention. Even if I'd invited a friend over when I'd just bought a new record, we would really listen, though maybe we'd also smoke a joint.

I devoted so many hours to the ritual of just listening. I don't do that as much anymore. A few years ago, feeling ashamed that I knew next to nothing about classical music, I took a couple of night courses to learn about it. The textbook talked about the need for "active listening." I understood why, and I heard so much more when I listened actively, but I really did it only because I was studying for the exams. Then I went back to my old habits—passive listening. I'm now too hyper and distracted to just sit and do one thing. And the technology makes that ritual of really listening seem anachronistic: the idea of inviting someone over to hear songs you just downloaded is laughable. Even downloading is starting to feel old-fashioned as people turn to YouTube and streaming services such as Apple Music and Spotify.

But passive listening risks turning music into aural wallpaper. Along with producing several great albums (including some by the Talking Heads), Brian Eno has devoted much of his innovative career to developing ambient music. Contemplating Eno's influence on music over more than four decades, Sasha Frere-Jones writes in *The New Yorker* that the *Ambient 1: Music for Airports* album is "too beautiful to ignore" and so it's a failure as ambient music. "But, in some ways, history and technology have accomplished what Eno did not," he goes on. "With the disappearance of the central home stereo, and the rise of earbuds, MP3s, and the mobile, around-the-clock work cycle, music is now used, more often than not, as background music."

If you ever find yourself in a room full of music psych-ologists and start to get bored, you can always liven things up by blurting out "auditory cheesecake." This is the advice Peter Pfordresher, a University of Buffalo psychologist, gave me. That's because there's no agreement on why humans developed the capacity for music. And a little uncertainty is all academics need for a food fight.

Music stimulates the ventral tegmental area (VTA), the pleasure centre that produces the chemical messenger dopamine and is linked to reward and motivation. The VTA, which is also turned on by chocolate, cocaine, and lust, triggers emotional responses that are separate from our intellectual ones. Except for a small percentage of people who suffer from specific musical anhedonia, an inability to derive pleasure from music, almost everybody enjoys listening to tunes. So it's no surprise that all existing cultures make music. Humans dig it. A lot.

All the extinct cultures we're aware of made music, too. In 2008, researchers found a bone flute and two fragments of ivory flutes in a cave in the south of Germany. They probably date from about 35,000 years ago, during the Upper Paleolithic period. They may be the oldest musical instruments ever found, now that archaeologists believe the Divje Babe flute, a 43,000-year-old cave bear femur discovered in Slovenia in 1995, was really just bones chewed up by hyenas. Surely, though, music making didn't begin with flutes — we likely started with percussion instruments. Or maybe the human voice. But we don't know which came first: speech or song or if they developed together.

That's where the delightful but divisive term "auditory cheesecake" comes in. Steven Pinker didn't drop it as much as detonate it in his 1997 book, *How the Mind Works*. A cognitive scientist and linguist, he believes language served an evolutionary purpose because it gave humans an advantage over other animals and over the environment. That was crucial, since humans are a little wimpy compared to some other animals. Music, on the other hand, is an enigma—and biologically useless.

"Music appears to be a pure pleasure technology, a cocktail of recreational drugs that we ingest through the ear to stimulate a mass of pleasure circuits at once," he writes. Sure sounds like fun to me. But that doesn't mean it played any evolutionary role. "I suspect that music is auditory cheesecake," writes Pinker, who calls it "an exquisite confection" that tickles several parts of our brains. By this argument, music is a by-product of language. It's enjoyable, but not crucial, just as cheesecake is enjoyable—tantalizing, even—but not an essential part of a balanced diet, despite our evolutionary craving for sugars and fats, which our ancestors sometimes had trouble getting enough of. In other words, if we didn't have music, humans would survive. Indeed, writes Pinker, "compared with language, vision, social reasoning, and physical know-how, music could vanish from our species and the rest of our lifestyle would be virtually unchanged."

Pinker devotes just eleven pages of his massive tome to music. And his choice of an empty-calorie dessert for his analogy, rather than a higher-brow but equally tasty food, suggests he might have been intentionally trolling his colleagues. Some of them agree with him, though. Dan Sperber, a French cognitive and social scientist, dismisses music as "an evolutionary parasite." American psychologist Gary Marcus, author of *Guitar Zero: The New Musician and the Science of Learning*, has Pinker's back, too. (No surprise, given that he studied under Pinker.) "To an alien scientist, music—and our desire to create it—might be one of the most puzzling aspects of humanity," writes Marcus.

Although he notes that the ability to hear pitch well would be helpful when moving around and avoiding danger, as either predator or prey, he makes a strong case against the evolutionary role of music. "What advocates of music as an evolved instinct often forget," he contends, "is that the music we see now—and that seems so compelling to us—is at least as much a production of *cultural* selection as it is of natural selection." By this argument, music is a cultural technology, developed over thousands and thousands of years, not something stamped on our genome.

While many regions of the human brain—including the temporal lobe of the cerebral cortex, the amygdala, and Broca's area—are active when we create or listen to music, the cheesecake gang is quick to point out that no region is dedicated to music. Instead, like reading, music takes advantage of several other parts with their own functions.

But the other side has its arguments and its own star witnesses, from Charles Darwin to high-profile contemporary thinkers. While Pinker grew up in Montreal and now teaches at Harvard University, Daniel Levitin was born in San Francisco and is now a professor of psychology and behavioural neuroscience at McGill University. Along the way, he worked as a session musician, sound engineer, and record producer. In *This Is Your Brain on Music: The Science of Human Obsession*, he admits that when someone of Pinker's stature issues a challenge such as "auditory cheesecake," it's a jolt to the scientific community. People re-examined positions they'd never bothered to question. "Pinker got us thinking," writes Levitin, though in his case, the reevaluation didn't lead to a change of mind. He's not alone: English archaeologist Steven Mithen also took umbrage at Pinker's suggestion and one chapter in Mithen's 2005 book, *The Singing Neanderthals: The Origins of Music, Language, Mind, and Body*, is called "More than cheesecake?" (Spoiler alert: yes.)

The pro-evolutionary crowd is armed with no shortage of theories to fling at the other side. Starting with sex, naturally. In *The Descent of Man, and Selection in Relation to Sex*, Charles Darwin writes, "I conclude that musical notes and rhythm were first acquired by the male or female progenitors of mankind for the sake of charming the opposite sex." The idea is that the more musical the man, the more mates he attracts. Every teenage guy who picks up a guitar tests this theory, with admittedly mixed results. Birds and whales use song in mating (and perhaps other reasons, including marking territory), though some experts contend this is more analogy than direct connection.

Levitin argues that the ability to sing and dance — which have only recently been seen as two separate things — indicated stamina and good physical and mental health to potential mates. More than that, musical ability is like the peacock's tail. "The colourful tail," he writes, "signals that the healthy peacock has metabolism to waste, he is so fit, so together, so wealthy (in terms of resources) that he has extra resources to put into something that is purely for display and aesthetic purposes."

The flaw in the mating and dating argument is that it doesn't explain the Ella Fitzgeralds, the Aretha Franklins, and the Neko Cases. Despite systemic sexism in the music industry, men do not have any innate musical advantage over women. In fact, music might be even more important to women. My mother may not have sung to me, but across almost all cultures mothers sing to their children. For most people, that's the first music they hear. Mothers use music to soothe their kids with lullabies, engage them with nursery rhymes, and teach them with jingles. And the sing-songy pattern that adults use when talking to babies, a musical speech called motherese, connects with infants more than normal speech or singing.

If the evolutionary reason wasn't sex, then maybe it was violence. Chimpanzees, much closer to our genetic home than birds or whales, defend turf and intimidate others with pant hoots, sometimes accompanied by drumming. Anthropologist Edward H. Hagen and theoretical biologist Peter Hammerstein argue that to proclaim territory, boast about their strength and number, form alliances, and express strategic emotional states such as anger, joy, and sadness, Neanderthals and other human ancestors "needed a signal that closely resembles music: a loud, group-specific, emotionally engaging chorus of highly synchronized sounds performed by group members who had practiced together for weeks, months or years."

One of the differences between speech and song is the rhythmic pattern. When we listen to something with a regular rhythmic pattern, humans tend to synchronize with that pattern. Even when we're at a classical concert and we're supposed to stay still, our brains are dancing with the music because our brainwaves are synchronizing with the music. Our idea of polite and docile behaviour at the symphony is a relatively recent social convention. But it's a powerful one, as Tom Morris, the artistic director of

the Bristol Old Vic in England, discovered when he tried
to revive the old ways. At a 2014 performance of Handel's
Messiah, he encouraged the audience to "clap or whoop
when you like, and no shushing other people." He even
invited them to create a mosh pit. But when a Royal
Society research fellow attempted to crowd surf during the
"Hallelujah" chorus, outraged patrons ejected him. Morris
claimed the last time something like this had happened at
a classical concert was back in the eighteenth century.

Because singing also helps us synchronize with other
people, it creates social ties. Early humans who sang and
played instruments together had a better chance of sur-
vival. A fan of this social cohesion argument, Levitin notes
that possible benefits of ancient campfire singing included
staying awake, scaring off predators, and nurturing co-
ordination and co-operation. For society to work, we need
connections with each other and music is one way to build
them.

Even without other people, we like to groove to our
favourite tunes. It gets us high when we do. Given that
our emotions are a control system for our bodies and our
minds, Mithen doubts that we "would be so easily and profoundly stirred by music if it were no more than a recent
human intervention. And neither would our bodies, as they
are when we automatically begin tapping our fingers and
toes while listening to music."

If you're still undecided about how essential music is
to humans, maybe Tom Chau can convince you. The boy
who liked to take things apart — "all kinds of things, radios,
cassette players, various kinds of toys, probably some

household appliances, too" — grew up to study biomedical engineering. But in 1999, after the birth of his first son, he left his job as a technical consultant at IBM to work, for half the salary, at helping kids. He started the PRISM (Paediatric Rehabilitation Intelligent Systems Multidisciplinary) Lab at Holland Bloorview Kids Rehabilitation Hospital in Toronto. Chau's thirty-member team of rehab engineers, therapists, and graduate students have one goal: liberating severely disabled children. And because of inventions such as the Virtual Music Instrument, he's earned a reputation as a world leader in technology for people with disabilities.

When he first started working on the VMI, Chau was thinking about movement, not melody. To encourage severely disabled children to be less passive and expand their range of motion, he wanted a way to reward them for waving an arm or a hand or whatever body part they could move. He and his team took a computer, a television, and a camera, then wrote software to create a device that plays different notes when a body part moves across coloured shapes on the TV screen. If the kids enjoyed making some fun sounds, that would be good, too.

One day, music therapist Andrea Lamont, who helped develop the VMI, showed a soft-spoken and mild-mannered eight-year-old boy with muscular dystrophy how an early prototype worked by playing "Twinkle, Twinkle, Little Star." Although the boy had no fine motor control and his fingers were chronically clenched, which made holding an instrument impossible, he played a note-perfect rendition of the song. The boy's parents had no idea their son had such musical talent. Chau watched their chins drop in disbelief. This was an emotional moment for everyone in the room—Chau experiences a lot of those in his job—but it also dawned on him that his invention could be a lot more powerful than he'd ever imagined. He just didn't know how powerful.

Since the first version of the Virtual Music Instrument in 2003, it has given children with disabilities a chance to engage in leisure activity while encouraging exploration and emotional expression. More than that, it lets kids who couldn't hold, let alone play, an instrument make music and helps them develop social skills. They can also play with able-bodied friends and family, free of psychological barriers. In 2010, the VMI won a da Vinci Award for innovative developments in assistive technology from the U.S.–based National Multiple Sclerosis Society.

Traditionally, locked-in syndrome refers to someone who, perhaps after a traumatic brain injury, is aware yet unable to move or communicate. But Chau sees a lot of people who are essentially locked in because of their disability. Music can be instrumental in unlocking them. A few years ago, his team worked with four adolescents who had no understanding of cause and effect. Trapped in their own bodies and believing that nothing they did had any effect on the world, they had developed a learned helplessness. Music therapists used the VMI to teach the teens that if they moved a specific part of their body, they could keep the music playing.

By the end of the study, two of the four had a clear understanding of cause and effect. In fact, one of the girls started moving her shoulder as soon as someone wheeled her into the room with the vmi. She'd never shown any comprehension of her environment before, but suddenly the ability to do something functional, such as activate a switch, became a possibility. "Music provides a medium of communication and self-expression that transcends cultural barriers and language barriers," said Chau. "And when you talk about kids who are non-verbal, essentially there is a language barrier. They speak a language that we don't understand, and they may be speaking that language through their physiology, through their facial expressions, through their movements. Music bridges the gap."

The lab has continued developing the technology over the years. The researchers have added dozens of instruments, everything from honky-tonk piano to bagpipes. In 2011, Eric Wan, one of Chau's grad students, used the vmi to play Pachelbel's Canon in D Major with the Montreal Chamber Orchestra at Place des Arts. Wan had played the violin before an adverse reaction to a measles vaccine left him paralyzed from the shoulders down at the age of eighteen. By moving his head, he was able to play pre-recorded violin segments—and realize his boyhood dream of bringing down the house in a prestigious concert hall.

The beauty of the vmi is the feedback it offers. Children can hear the tones they produce—and see themselves activating the sound. "That cause and effect is really essential," Lamont said. "Once we get them hooked, there's that huge potential for learning."

Meanwhile, the boy who played "Twinkle, Twinkle, Little Star" used the vmi and the Suzuki method, which teaches children to play an instrument by breaking the skill into small steps in a positive environment, to continue developing his musical ability. That helped change the way he thought about himself. One of the first things he said to Lamont when she met him was "I am disabled." A few years later, she overheard him introduce himself to a reporter by saying "I am a musician."

I may not be a musician and I may not have been born with a natural ability to sing well, but that doesn't necessarily mean I wasn't born musical on some level. This is another bun fight among the academics. Steven Mithen argues that we're all born with an inherent appreciation for music. In addition, we all have some innate knowledge of music: we're not born knowing all the complexities that grown-up brains can grasp, but day-old infants can detect a beat in music and, after three months, they can detect pitch changes. Gary Marcus, on the other hand, writes: "To the degree that we ultimately become musical, it is because we have the capacity to slowly and laboriously tune broad ensembles of neural circuitry over time, through deliberate practice, and not because the circuitry of music is all there from the outset."

As such debates drone on, it's no wonder some cognitive psychologists have an allergic reaction to evolutionary psychology. What frustrates Pfordresher is that the evolutionary adaptation debate tends to fall back on definitions of something that keeps changing. "It's very hard to define music," he said, "in part because people in the business of making music often go to great efforts to test the boundaries of what we call music."

Pinker's analysis focuses on instrumental music and avoids singing. That streamlines and simplifies the argument. "The problem, though, is that singing may well have been where music started," according to Pfordresher. Indeed, what we now call speaking might not have been that easily distinguishable from what we now call singing. Even today, speaking and singing can be hard to distinguish. Early speech in babies and children is often song-like— a baby's cooing is more like singing than speaking—while motherese is a cross between singing and speaking. And even Pinker notes, "Some singers slip into 'talking on pitch' instead of carrying the melody, like Bob Dylan, Lou Reed, and Rex Harrison in *My Fair Lady*. They sound halfway between animated raconteurs and tone-deaf singers."

Mithen doesn't believe language came first. He doesn't believe music came first. And he doesn't believe they developed together. Instead, he argues, "There was a single precursor for both music and language: a communication system that had the characteristics that are now shared by music and language, but that split into two systems at some date in our evolutionary history."

That we'll probably never know—fossils, after all, aren't much help when it comes to answering questions about early human cognitive capacity—has only escalated the food fight rather than squelching it.

Some people want to believe that if music played an evolutionary role, that gives it more legitimacy today, especially as an artistic endeavour. That was my gut reaction, but eventually I had to admit that it just doesn't matter. All you have to do is look around and you'll see how common music is, how popular it is, and how significant it is in our lives. That should be all the legitimacy anyone needs. As Sean Hutchins, director of research at Toronto-based Royal Conservatory of Music told me, "I don't think you really need to go hunting for explanations in the deep evolutionary past to make music a meaningful part of people's lives."

No matter how humans came to love it, song clearly means so much to us. Many folks devote their lives to making music, even if it's only as an avocation — and many others wish they could. Including me.

"For the Sake of the Song"

Partway through my first lesson, Micah Barnes asked me how I was feeling, psychologically and emotionally, as I was doing the exercises. "Trying to hit pitches," he said, "can make us nauseous if we're uncomfortable."

Fortunately, he was the opposite of the stereotypical Teutonic music teacher who sternly delights in scolding and wrapping knuckles. I appreciated his encouraging style. From the beginning, he seemed a perfect fit for me: he's into the mental, emotional, and psychological side of singing, not just the technical aspects. Barnes was as much therapist as singing coach.

More than that, though, he had assured me I was trainable. He may not have realized what he was getting into, but throughout my lessons he kept repeating that mantra. I sometimes wondered if that was for my benefit or his. Did he believe that if he said it often enough, it would come true?

"I guess I could be bummed that I'm getting only one out of the three notes right, or I could be like, well, I'm getting one right," I said. "One is better than zero."

I tried to explain how clued out I really was. I reminded him of the first time we met to see if he'd be willing to take me on. He'd tested me and said I was singing harmony naturally (though in a future lesson he will admit that even my unconscious harmony singing is a little out of tune). But I had no idea what that meant; I thought singing harmony meant singing exactly the same notes as the other person.

"So," I admitted, "that's just how much of a blank slate you're working with here."

Ask kindergarteners if they can sing and they'll all put their hands up; ask high school students, and only a few will. So it's learned behavior. Or, rather, it's behaviour caused by a lack of learning. Drawing is the same. Stephen Zeifman, a former high school teacher who now runs an art school in Port Rexton, Newfoundland, says that as young kids, we all know we can draw—hell, our parents even hang our stick-figure masterpieces on fridges—but at some point we decide we're no good at it. Probably because we see someone who's better than we are, which is an understandable, but silly, response. (After all, you don't have to be the best player on a team to enjoy playing a sport.) Regardless of why we've stopped, Zeifman says he can teach just about everyone to draw. And he can identify the tiny minority who can't learn with a few quick and simple tests.

When it comes to the arts, especially singing, we

separate the talented from the herd early and decisively. We don't invite kids to try soccer and then after one practice, invite the best ones to come back and tell the rest to find some other way to have fun. But we do that with singing.

The classic story, one I've heard over and over again — and one even Barnes has heard from some of his students — is of kids told to mouth the words instead of singing out loud. My unscientific survey suggests that people often get the devastating news about their singing somewhere between the ages of eight and ten, though sometimes it comes later. Some humiliations are more public than others. When Lyle Lovett, the great American country singer-songwriter, was going to high school in small-town Texas, the students all had to line up against the lockers. As a teacher played piano and everyone sang, the principal walked down the hall, listening for pitch. The good singers, including Lovett, stayed in line. The bad singers didn't — the principal grabbed them by the shoulder and dragged them to study hall.

Perhaps the most bizarre, and certainly the creepiest, story comes from former radio producer David Shannon, who attended Selwyn House, a private boys' school in Montreal. The place fancied itself as terribly British, and the choir was part of the curriculum in the junior school, which started in grade one. At the beginning of the year, every boy in the early grades had to sing a solo, an Anglican hymn. Upon hearing the boys sing, the choir teacher separated them into three groups: Archangels, Angels, and Devils. The Archangels, who had the best boy soprano voices, sat closest to the piano in the choir room. Out of a class of fifteen, three or four might achieve this exalted status. The Angels, those with decent voices, sat behind the Archangels. Three or four Devils sat several rows behind them, at the back of the class, where they were essentially ignored. The teacher tolerated their singing in class but didn't encourage them. And for performances at important school assemblies or events that parents attended, they were to just move their lips.

The teacher was British, had barely a hair on his head, and wore long blazers over trousers that he hitched up so high that the belt seemed to sit just under his nipples. He'd been around forever, and the choir program was his. He did it his way. Knowing that boys like to compete, he tried to engage them with little contests. As they sat at their desks with hymn books closed, he'd announce, "Hymn 124," and the first boy who could open his book and find the right one would exclaim, "Got it, sir."

"You get an egg," the master would say. Then he'd lift the side of his long blazer so the winner could reach into the deep pocket of the man's baggy trousers and fish through the keys and coins and other crap to find a small candy. To six- or seven-year-olds, these seemed liked magical candies. (Years later, Shannon realized they were just Scotch mints.) Upon grasping the prize, the boy would declare, "I've got it, sir," and pull out his reward.

The best singer of the day would also get a chance to dig for an egg. But the Devils never got candies. "The Archangels were puffed up, the Angels felt good enough, but the Devils were shunned," Shannon explained. "In sports, or art or math, we all knew some boys were better than others. But nobody was singled out as being a lost cause and not worthy of attention to improve them or even sympathy. Everyone has heard that standard, 'Oh, I was the last one picked for teams,' but at least you were picked. You were on the team. The Devils were completely shunned."

Shannon learned what that felt like after a twist of fate otherwise known as puberty. Like two of his four brothers, he'd been an Archangel as a small boy. Even as a tween, his voice was so good that a nearby Anglican church wanted him for its choir, one that toured internationally—and paid. The problem: he was a Catholic altar boy, which paid only an occasional tip after a funeral. This created a dilemma for his parents, and as they negotiated with the two churches, Shannon's voice began to change and went so awry that he could no longer even carry a tune. "That was traumatic," he said. "It took away an opportunity and it took away status and it was all based on something I had no control over."

The two churches stopped haggling over him and he found himself treated like a Devil. He never really sang again. "I like to sing and I sing a lot to myself and I can do faux Broadway show tunes singing with people—really loud and basically talk-singing—but I don't think I could sing anymore."

I asked clinical psychologist Alex Russell for his thoughts on children and music. Before Russell became a psychologist, he wanted to be a composer and he'd even started his university career studying music. A few years ago, I helped him write a book on parenting, and I knew he had strong opinions about the role of play in childhood development.

He started by reiterating what many others have told me: music creates bonds between people. As someone who doesn't make music with other people, I understood that only on a theoretical level. So he put it in terms I could understand: "We can say, I love my wife or I love my children, but when we're doing music, we're actually connecting emotionally, like we do when we have sex with one person. I can talk about my lover—I love my lover—but when I'm actually making love to my lover I'm experiencing a connection, a lived connection." There are other ways to have this experience, such as a profound, intimate conversation. But making music together may be the best way for two or more people to connect.

I saw thousands of people connect when I travelled to Chicago in 1994. My friends and I had flown to the Windy City to watch a hockey game between the Chicago Blackhawks and the Toronto Maple Leafs. Well, really, we went there for the singing of "The Star-Spangled Banner."

It was the most powerful rendition of an anthem I have ever witnessed. Two events—one planned, one tragic—only served to heighten the emotions in the building that night. First, it was the last regular season game to be played at the sixty-five-year-old arena before it gave way to a new one designed to maximize revenue. Second, Wayne Messmer, the man who sang the anthems at the stadium, was fighting for his life in hospital. A fifteen-year-old with a handgun had shot him in the neck five days earlier.

Announcer Pat Foley read a statement from Wayne Messmer and then introduced Kathleen Messmer, who stood where her husband should have been: beside an honour guard in the organ box. Foley went on to ask the crowd to "raise the roof" for the injured singer during the taped anthems. After "O Canada"—which more people in Chicago sang than do at games in Toronto—it was time for "The Star-Spangled Banner."

As a schoolboy, I was taught to stand at attention during a national anthem. In Chicago Stadium, however, people did it differently. They sang, clapped, cheered, whistled, lit sparklers, punched balloons in the air, and generally worked themselves into ecstasy. It's a tradition that had begun about a decade earlier and then took on a life of its own during the first Gulf War.

From the first note, the noise was deafening and the emotion high. But what astonished me was how it built. With each bar, the thunder defied logic by growing louder. Meanwhile, the eyes of the working-class Chicago men—not to mention the odd Canadian visitor—grew misty.

Cognitive psychologist Frank Russo's working hypothesis, which reflects his ideas about music as movement, is that when we sing together we move together in perfect synchrony. This allows us to feel emotion in the music and to feel connected as a group, opening us up to trust, understand, and like the people we are singing with. The reason we sing national anthems isn't just patriotism (or jingoism, depending on your point of view), it's because singing them makes people feel they belong. "It's a shame that we can't get a new song every month," said Russo, "because the national anthem loses some of its flavour after a while."

Andrew Cash, who was in the respected Toronto punk band L'Étranger and then enjoyed a successful solo career, had a similar experience while travelling through India in the 1980s. Cash, who later served as a member of Parliament, found himself on a packed train at 7 a.m. He was the only Westerner in a car full of young men commuting to a textile factory outside Mumbai. Seeing his guitar, they asked him to sing. He declined. Surprised, they pressed him again, and when that didn't work, one man said, "Okay, we'll sing first."

The workers traded songs for half an hour until Cash finally pulled out his guitar. The other passengers didn't know "Monday Morning on the Move," a song Cash had written about going to work, and they spoke little English. But they sang along anyway, making up the words as they went. Years later, Cash still marvels at how comfortable the young men on the train were singing out loud, in public, and at such an early hour. "It was no big deal," he told me, "just something they did."

The loss of play — musical or otherwise — concerns Russell, and he's seen it decline even in the last generation. We're not giving children a chance to play the way they have for thousands and thousands of years of human civilization. Because we're too focused on making sure kids develop skills and play according to sets of rules, we aren't allowing them to go out and muck about and create their own play. Music is the perfect example of this. "We're not giving human beings a chance to just make a whole lot of fucking noise together and it doesn't matter how good or bad," he said. "It's the doing of it that matters."

And yet he understands why we're self-conscious about singing in public, especially people who were told at a young age that they can't or shouldn't sing. "There's something about singing that feels deeply revealing of the self," he said, noting that it's not the same when you're playing an instrument or even clapping. "It's almost like taking your clothes off. You're revealing something very personal about yourself when you sing."

If we could tap into the minds of typical kids today to get a sense of how they understand music and what they think music is, Russell figures we'd find they believe it's the highly refined, polished sound they hear in YouTube videos or on their phones or iPods. That's what music is to them; that's how it's supposed to sound. So the gulf between the music makers and the music consumers continues to grow. And people will always be tempted to evaluate sound for how good or bad it is, rather than simply appreciating the making of it.

Our emphasis on consuming rather than creating leads children to say, I'm not a musician, I'm not good at that, or I can't sing. And then, naturally, they don't participate. We've professionalized music so much that we discourage people from making it for fun. For generations, people sang whenever they were together. Families gathered around the piano after dinner. Friends sat in kitchens, singing and playing fiddles. People made long road trips seem shorter with singalongs. The cynical might point out that there was nothing else to do back then, and they wouldn't necessarily be wrong, but singing together was a normal part of life for a long time. And then, one new technology after another began to change that: player pianos, phonographs, and radios. When people started buying TVs in the 1950s, the submission to technology became increasingly unavoidable. Instead of entertaining ourselves, we began to let others do it for us.

Each new technology helped sell more music. The radio helped turn the likes of Frank Sinatra into stars, and despite fears television would kill the radio stars, it was essential to the hysteria over Elvis Presley and, later, Beatlemania. But recorded music had another unintended consequence: suddenly people who made music had to compare themselves to the best in the world. Someone who might have been a renowned pianist or fiddler in a town might seem just okay compared to what was available on recordings. In a society built around specialization, some people naturally assumed it wasn't even worth trying. So we increasingly became consumers instead of creators. After all, why make music when the Chairman of the Board or the Fab Four could do it so much better?

Meanwhile, many of us stopped going to church. That was one place people regularly congregated and sang together, even after the TV replaced singing around the piano as a post-dinner activity. As a small boy with no interest in religion, I found the chance to sing in my father's Presbyterian church the most tolerable part of the service. Not that I particularly liked most of the hymns, but at least I could stand up and do something. And I didn't care if I was a bad singer back then.

Our approach to music education hasn't helped. The emphasis went from playing to listening, which had a completely predictable effect on the number of people creating their own music. Nor was it wise to focus on classical music when kids just wanted to listen to their own generation's soundtrack. The aim, notes David Byrne in *How Music Works*, was to convince young people "to appreciate the superiority of a certain kind of music over what some declared to be coarser, more popular forms." He also argues that funding high art discourages amateurism: "It can often seem that those in power don't want us to enjoy making things for ourselves—they'd prefer to establish a cultural hierarchy that devalues our amateur efforts and encourages consumption rather than creation."

So by the end of the '60s, a strange thing had happened to our relationship with music. Popular genres provided the soundtrack to a cultural revolution and yet, even as hundreds of thousands gathered at festivals such as Woodstock, we were losing our connection to music as something we created together. Instead, we consumed together.

Soon we were doing more and more music listening on our own. By the early '80s, the Sony Walkman and its knockoffs gave us a chance to be alone with our tunes even in crowded places. Later, Apple's iPod accelerated this phenomenon by allowing us to carry massive music libraries wherever we go.

Music also changed. Many popular genres were perfectly suited for amateurs. The simple guitar and vocals of folk, for example, was ideal for campfire sessions or jamming in rooms thick with marijuana smoke. Later, the do-it-yourself ethos of punk—in part a reaction to the excesses of progressive rock, some of which required a lot of technology or even an orchestra—encouraged the formation of countless garage bands. Earlier, the proto-punk band The Velvet Underground had a similar effect. In a 1982 interview with *Musician* magazine, Brian Eno famously talked about the low sales of the band's debut release, "I was talking to Lou Reed the other day and he said that the first Velvet Underground record sold 30,000 copies in the first five years. The sales have picked up in the past few years, but I mean, that record was such an important record for so many people. I think everyone who bought one of those 30,000 copies started a band!"

Melody started to wane with the rise of rock 'n' roll. Beat-heavy styles such as rap and electronic, two of today's dominant popular genres, rely even less on it. The further music gets away from melody, the harder it is to reproduce without technology. I remember when a great pop tune was "Tempted" by Squeeze. But now the most popular pop songs—most often sung by a beautiful woman who can dance well—are heavily produced and Auto-Tuned, while rap and electronic depend on studio technology and sampling. Inexpensive software means that kids with the musical and technical skills can replicate this music on computers in their bedrooms, but not around a campfire.

Meanwhile, fragmentation in the music industry means we have a smaller common stock of songs for campfires, anyway. As a little kid, I heard music on Top 40 AM radio and, to a lesser extent, on television. Everyone knew the same songs. In the '70s, I started listening to FM radio and reading music magazines, expanding my tastes beyond the mainstream, but not abandoning it. By the end of the decade, Toronto had three FM rock stations that were distinct enough that you could tell a lot about people by which one they listened to. Later, the Internet splintered us in ways we couldn't have imagined when I was a boy. It still shocks me when I realize that some band I've never heard of plays a huge venue such as the Air Canada Centre, Toronto's hockey arena. Ethnomusicologist Gerry McGoldrick remembers being on a party bus in the '80s, and the only songs everyone knew were the theme songs from old TV shows such as *The Flintstones* and *Gilligan's Island*. Group sings would be even tougher to pull off today because we don't even watch the same programs anymore.

True, we don't need to sing to enjoy music any more than we need to paint to enjoy Vermeer or Monet or Basquiat. But we've professionalized it so much that few but the talented and trained now have the confidence to sing in public, unless we count alcohol-fuelled karaoke and some of the more deluded entrants on reality TV shows.

The ability of amateurs to sing a number is just about the last thing on the minds of North American musicians and producers when they record a song. In Japan, though, singability is often the goal, according to McGoldrick. Music producers there know that a track that's ideal for karaoke has a better chance of becoming a hit.

After first becoming popular in Japan following the invention of the karaoke machine in 1971, the pastime eventually spread around the world. And what seemed like a sure-to-be-short-lived fad—destined to go the way of riding mechanical bulls in bars—endured. It remains hugely popular in Japan, though rather than perform in front of lots of strangers, the preference there is to do it with friends in smaller rooms called karaoke boxes. When McGoldrick taught in a Japanese high school, he knew it was a singing culture but he was amazed at the level of music literacy. "They could all goof around on the piano," he said, "or at least read the notation."

Karaoke certainly has many, many fans in North America, and in a society that loves competition and profit, not necessarily in that order, the logical extension of it was *American Idol.* This show, based on Britain's *Pop Idol,* was one of the most watched televisions programs in the United States from 2002 to 2016. Its come-on was that anyone could be a star. Just line up, try out, be discovered—and you too will land a lucrative recording contract. Sounds democratic, doesn't it? To lots of people, it did. But to others, rather than democratizing singing and encouraging everyone to do it, *American Idol* and its ilk ended up reinforcing the idea that only good singers should do it and everyone else should shut up.

To learn more about these programs, I visited Sue Brophey at Toronto's Insight Productions, the company behind the six seasons of *Canadian Idol,* the Great White North's version of the phenomenon. Using Calgary, where about 1,200 people would line up for the open call each year, as an example, Brophey, who had been supervising producer of the show, explained how the auditions worked.

She was one of half a dozen producers who would see contestants in groups of five or six. Each person would sing a
verse and a chorus of a song he or she had chosen. Then the
next group would come in. About 300 singers would go on
to the next round, singing individually for a producer, who
would also ask questions to learn a bit about their personal
stories. About 150 singers would make it to the third round,
on a different day, and perform before the celebrity judges.
The auditions that appeared in the first two episodes of each
season came from the third round.

Brophey, who admits she's a bit of a ham and even
sang with the house band to pump up the audience before
Canadian Idol tapings, was surprised at how many bad singers tried out for the show. A few, inevitably, were doing
it as a lark and sucking was the point, but many didn't
realize how inept they were. She figured more than half
shouldn't have tried out. Some couldn't carry a tune. Others
could, but their technique was poor (nasally singers and
bad breathers, for example). Still others could do it well
enough but were just too boring. "Being a singer isn't just
about singing perfectly in tune, it's also about emoting, it's
about feeling the song and communicating the feeling you
have about the song to the audience," she said. "You had to
have the whole package." Material was also a problem for a
number of contestants. Some teenagers picked songs about
life experiences they couldn't possibly have gone through,
and it was clear they didn't relate to the lyrics. "You have
to connect with the audience and how you do that is to
actually understand what you're singing."

If she heard 175 people a day at open calls, ten of them

would be good singers and three might be real standouts. A quarter might make for good television, though, and that's why the audition shows featured the good, the bad, and the eccentric. "We got huge numbers on those audition shows because people like to see a train wreck," she admitted. "They want to see some great singers but they also love seeing the bad singers." Those early shows were particularly popular with younger viewers—sarcastic college kids and so on—but when the only contestants left were good singers, a slightly more mature, but still enviably young, demographic tuned in as they became invested in the characters.

I asked her if she thought these shows encouraged singing. "Oh, for sure. I think when people watch the audition shows, they don't ever go, 'That might happen to me,' because they think they are good singers," she said. "There are people who think they're really good and they're not."

At one of my lessons, Barnes and I talked about my goal of singing in public and my fear of being laughed at. He asked if I really thought people would laugh at me and I pointed to *American Idol*. That sent him on a riff about it. "Yeah, but that's a TV reality show built to entertain. So you've fallen prey to the same old entertainment capitalism that everyone has. Watching the circus of who's a good singer and who's a bad singer and how to create viewer interest. Somebody bad so we can laugh is part of it, and so is discovering somebody good. Who wants to look at a car crash? Who wants to see a princess? It's radically altered what singing is."

"For better or for worse?" I asked.

"For way worse. Now we think of it as a high-flying

trapeze act. *American Idol* is like the Romans having gladi-
ators so we can see blood. It's win–lose, success–fail, human
drama. The singing is just not even part of it, really." He
pointed out that the show had failed to discover any artists
that were of any interest to him, before admitting, "Okay,
we have Kelly Clarkson's Christmas album, but that's
because my boyfriend's gay."

But his joke wasn't the end of his rant. "I'm very, very
down on that show," he said. "As a coach, I slave every day
against the concept of you being a bad singer going on
that show so we can have the drama of *that guy can't sing*.
It's so counter to what I'm doing as a coach, trying to find
authentic voices who are artists."

My voice may be authentic, but obviously I'm no artist.
Still, Barnes never suggested I give up. We kept at it, and
my education continued as he patiently answered my stupid
questions. At one point, he said, "You've got the note but
not the pitch."

Later, I realized this was just his encouraging way of saying I was a little out of tune. But at the time I was confused
and asked, "So the note is like the C, G, A, whatever . . . and
the pitch is A-flat?"

"The pitch is the same as a note. If I say you're out of
pitch, I'm saying you're not singing the note."

"So I shouldn't get confused about those terms."

"People use them differently. People will say someone
has good pitch or bad pitch or perfect pitch. God, wouldn't
that be nice? I don't have it. It's referring to your ability
to hit the notes. So pitch is your overall ability to hit the
notes."

Maybe, I began to realize, this was going to be even harder than I'd imagined. And learning the terminology was the least of my problems — I wanted to figure out exactly why I was a bad singer.

"The Harder They Come"

"What's interesting about teaching you is that I'm forced to get elemental about what music is made of," Barnes admitted before we started doing any exercises in our second lesson. "It's interesting to me because it's nuts and bolts."

I didn't have time to worry about that making me feel like a bit of a freak, because soon we were working on my breathing. He had me do some sighs — *haaahhhh* — but I wasn't doing them right.

"You're pushing air and that's where a lot of your stuff comes from. You're throwing yourself at it without being relaxed," he corrected me. "Your ear will change. Usually when someone has bad ears, they're grabbing for the sound instead of just letting the sound happen." He made me lie down on the black couch again so I could work on the mechanics of breathing properly. After a while, he let me stand and, happy enough with the results, said,

"That was good news! I'm just looking for inches, I don't need feet."

Next, I had to match his notes by humming while moving and twisting and scrunching my lips—manipulating them, he called it. He asked me to look in the mirror to make sure I was breathing from my diaphragm. The goal of this exercise was to feel my lips buzz, or resonate, when I hit the right note, matching the one he played on the piano. "Feel that buzziness on your lips?"

Barnes noticed that looking in the mirror calmed and improved my breathing. "Are you thinking you're doing this stupid, weird thing?"

"No," I said. "I'm just worried that I'm doing a good thing badly, stupidly."

"At this point your expectation should be that you're doing it badly."

"So I shouldn't be doing this in public."

"Not today," he admitted. "But this is going to add up, you watch. Now when you leave you're going to say, 'Oh, I'm hitting my pitches a bit easier.' That little bit of confidence will start to buoy you up. There are a lot of coaches whose job it is to make you think they're God and then tell you, 'No! That's not it!' My concept is warm bath water. If you're batting 50 percent, that's awesome. It doesn't mean that you're singing perfectly; it means that you're letting yourself try."

We continued with the exercises. "Let's pant, because we know how to do that one."

I panted: "*Ha. Ha. Ha.*"

"Tongue like a piece of meat," he instructed.

I panted with my tongue out.

"Excellent."

"There are so many things to think about: the note, the tongue, the breathing," I complained.

If singing was this complicated, I thought, it really is a wonder anyone can pull it off.

From an early age, talking comes pretty naturally to most of us. We open our mouths and out come words. Sure, we still have to build our vocabulary and our grammar, and maybe ease up on the ten-year-old boy's fascination with all things scatological, but talking is easy enough. Singing seems like it should be just as easy, and for some people, it is. But there's actually a lot going on in our bodies when we sing.

When we want to produce a note, we need to co-ordinate our lungs, our diaphragm, our throat, our mouth, our tongue, and our lips. And so much—from a change in lung pressure to the wrong shape of our mouth—can go wrong. We need to make all those elements work together when we talk as well, but speaking English and many other languages requires producing specific sounds, not specific pitches. So while we may tease someone who uses a funny pronunciation of a word, most of us are tolerant of people with accents. And we don't call people with deep, sexy voices bad talkers because they don't hit the high notes.

Sean Hutchins, of the Royal Conservatory, compares singing to tossing a baseball. We require complicated muscle co-ordination and timing to throw a ball with the right trajectory. Screw up any step in the process and the ball won't go anywhere near your friend's mitt. "The same type of thing is going on when you're singing," he said, "except it's all inside of you." That makes it much harder.

While we can watch someone throw a baseball to learn how to do it better—just as we can watch people play piano or guitar and see what they do with their fingers and hands—the steps necessary to sing a note are invisible. Watch singers and you will see their lips move. And perhaps you can see their bodies move as they take in air and let it out again. But you can't see what their diaphragms are doing or what their lungs are doing or even what their tongues are doing. Everything is hidden. Even if you used a laryngoscope to see these muscles and organs, you probably couldn't make much sense out of what you saw. According to Peter Pfordresher, "It's a pretty strange-looking image."

Growing up in Connecticut, Hutchins took music and voice lessons throughout high school. He kept at it during his undergrad and graduate years with some a capella, some barbershop, and some light opera. He even met his wife while doing Gilbert and Sullivan. Meanwhile, as a psychology student, he was intrigued by the relationship between perception and production: how we co-ordinate what we see and perceive in the world with what we do in the world. As it turned out, singing was ideal for studying that interaction because in order to sing together, singers must co-ordinate what they're producing with what they're hearing.

If you can't hit the right notes, you are, by definition, a bad singer. So the starting point for all good singing is the ability to perceive pitch, but it's more complicated than that.

After a music teacher said "I think everybody is deeply musical," the phrase stuck with Pfordresher and it's something he's wanted to explore through his research ever since. Believing it's a shame that so many people are unhappy with their singing, he wanted to understand the problems they have with it and to see if there was a way to help them. He thinks singing is as fascinating as it is difficult. Academic literature is full of studies, including his, about why some people can't imitate pitch correctly. "An almost more perplexing question is, Why is it that anybody can do it?" he said. "The fact that as many of us can do that as are able to do that is, to me, a kind of wondrous thing, a kind of mystifying thing."

He believes that when people say, Oh, that guy can't sing, they usually think they're assessing pitch. But they may be reacting to something completely different. When he talks to people about his job, they inevitably bring up American Idol and all the hopeless performers on the show. Many of them are, he'd agree, poor-pitch singers, but he figured there was something more going on so he decided to evaluate the program's most notorious contestant: William Hung.

After winning a talent contest at his Berkeley dorm, the civil engineering student auditioned for the third season of American Idol. Despite his obvious (and nerdily charming) enthusiasm, it did not go well. While Randy Jackson, one of the judges, chortled away, Simon Cowell, the famously nasty judge, stopped Hung's a cappella rendition of Ricky Martin's "She Bangs" early on and said, "You can't sing, you can't dance, so what do you want me to say?" After his performance ran in January 2004, it became a viral sensation and Hung became a celebrity—and even landed a record deal.

Curious about why everyone thought it was so terrible, Pfordresher compared Hung's rendition to the original. "Sure enough," he said, "William Hung's pitches were pretty much identical to Ricky Martin's pitches." He was singing the right notes, and he was doing it in one take without accompaniment and without digital editing, in a stressful situation. Not many people could do that. And still they laughed him off the stage.

Turns out, there's more to a good voice than pitch control. Timing, for one thing. Volume, for another: some songs call for soft, gentle vocals while some music demands it be sung loud. In addition, every singer has his or her own distinct sound, and some sounds work better for certain songs than for others. Luciano Pavarotti's rich tenor was no more likely to work well for an album of country and western covers than Johnny Cash's deep growl would have been appropriate for an opera.

Hung's problems started before he even hit the stage. "She Bangs" wasn't a wise choice—not because it's a silly song (though that's true, for sure), but because it's not well suited to an a cappella rendition. And, sad to say, Hung's accent didn't help. From many of the original British Invasion bands to more recent examples such as First Aid Kit—two Swedish sisters who come across like they grew up in the American Midwest—we're so used to artists from all over the world sounding as if they're from the United

States that anything else seems off. Then there's Hung's timing, which isn't great, and the thinness of his voice.

In some genres, not being a virtuosic singer doesn't preclude great success. In 1985, Canadian musicians gathered in a Toronto studio to record a benefit song called "Tears Are Not Enough." It joined the British "Do They Know It's Christmas?" and the American "We Are the World" as supergroup singles that raised money for Ethiopian famine relief. The song is hardly an enduring classic, but the list of musicians who recorded it remains impressive: Neil Young, Joni Mitchell, Gordon Lightfoot, Anne Murray, Liona Boyd, Oscar Peterson, and many others. After Young sang his line in the studio, producer David Foster told him, "Little flat on 'innocence,' but other than that, it was great. We'll go again." Young waited a beat or two and then deadpanned, "That's my sound, man."

That sound has certainly served him well for nearly six decades. Perhaps that's no surprise in rock music, but opera singers can sometimes be as far out of tune as shower singers. Divas have a certain sound that comes from the way they support their voices and the way they hold their facial muscles. "And then there's that little trump card in classical singing, the vibrato, which, frankly, makes the pitch that someone's singing somewhat ambiguous," said Pfordresher. So they can be slightly off and few of us will notice.

Even for good singers, getting really good takes work, as I learned from Kellie Walsh, who comes from what she calls a "traditional Newfoundland family." Her dad was the son of a fisherman who had a family of eighteen and, when she was growing up, Sunday dinners were at her grandparents', where there was always music: people played accordion, spoons, and guitar and they sang and danced.

As the artistic director of the Lady Cove Women's Choir and Shallaway: Newfoundland and Labrador Youth in Chorus as well as the founder and conductor emeritus of the Newman Sound Men's Choir, she's taken part in competitions around the world. The trick to singing well, she says, is to treat the body as an instrument rather than just thinking, If I open my mouth, sound will come out. Young singers often don't realize that a good voice comes from their diaphragm, not their throat. Once they start using more of their bodies, they produce a more pleasing sound.

The goal with a choir is to make the sound as "blended and beautiful and in tune as possible," but trying to instruct and correct up to sixty singers at once is, of course, more challenging than working with one person. And because singing takes place inside the body, which she can't show, Walsh uses metaphor and mental images. One of her techniques is to tell singers to think of the inside of the mouth as a clock. Sending the sound to twelve o'clock means directing their voices up to the centre of the roof of their mouth, or to three o'clock where the teeth are. "I'll say put that vowel right at twelve o'clock. Or if I want it to sound darker, I'll say put that vowel back at nine o'clock, to the back."

While Barnes spends a lot of time working on technique with his clients, he also finds that the hardest part of singing for most of them is trusting their instrument, which requires trusting themselves. "Singing is easy," he told me. "It's just hard to let it be simple." Our need to be good, our worrying about being good, and our eagerness to criticize ourselves complicates the act for us. We end up afraid to do it. "If you're full of doubt and judgement, then singing is indeed quite hard. To be worrying about your singing while you're singing means you're doing it wrong."

As a performer, Barnes knows that when he's relaxed, he sounds better and puts on a better show. He believes singing is emotional, spiritual, and physical — and you can't separate this trinity. He's met and worked with a lot of great singers and they tend to be really soulful people. "Gladys Knight is a great singer, and she was exactly like her voice: warm, relaxed, intimate, easygoing. Chaka Khan is a brilliant singer, and she was exactly like her voice: intense, radical, crazy. k. d. lang is a very relaxed person, a very self-assured person, a very in-touch-with-her-spiritual-self person. Really surrendered in a Buddhist sense." Still, he admitted, truly good singing is hard. "It takes an unusual amount of trust and faith to really fill the world with your sound."

Trust and faith aren't always enough to solve every-one's singing problems. Sometimes there are technical reasons our sound sucks. According to Hutchins, bad singing — or poor-pitch singing, as researchers tend to call it — has three main causes. The first is a motor control problem. This is a difficulty co-ordinating the necessary muscles in just the right way to produce an intended note. These people are like unco-ordinated athletes. "They still have arms, they still have the ability to pick things up to throw things," said Hutchins. "They just might not be able to throw them very accurately towards an intended target." These bad singers know what notes they're trying to hit, and they can hear when they miss. They account for about 20 percent of the population.

In most cases, training can help. Pfordresher played keyboards and rhythm guitar in high school rock bands, but he really wanted to be the lead singer. Whenever he tried, though, people told him not to do it again. Frustrated, he wondered what he was doing wrong. His pitch perception was fine and he was, technically, hitting the right notes. Once he realized he had a motor control problem and worked on it, his singing improved.

The second cause is what Hutchins calls a timbre translation problem. Every instrument or voice has a distinct sound, so matching pitches is like exchanging dollars and drachmas — some conversion required. While a C on a piano, a C on a saxophone, and a C sung by a person are all the same frequency, they don't sound exactly the same because of the different timbres of the three instruments. So to duplicate a piano tone, for example, your brain has to figure out what note it is on the piano and then convert it for your voice. That's why it's much easier for a singer to match a human note than one produced by an instrument. People with timbre translation problems may be able to sing a song just fine on their own, but will likely struggle singing with instruments or even other people. Sometimes they will have trouble with karaoke if they misperceive how their voice aligns with the backing track. But these singers, who make up about 35 percent of the population, aren't the ear-splittingly bad singers, said Hutchins. "It's the people with the motor problem and the straight perceptual problem who are really going to burn your ears off."

The perceptual problem he was referring to is the third and most serious, but least common, reason for bad singing. Tone deafness, or congenital amusia, is a neurological disorder that affects the way our brains process music leading to an inability to hear small differences in pitch. And when the only real cue is the sound and you can't hear it, or you can't hear it well enough, that won't help your singing.

You also need musical memory to be able to retrieve many notes in a complicated musical structure, another skill most amusics are poor at. These people usually can't hear what they're supposed to sing and couldn't reproduce the notes even if they knew which ones they were trying to hit. Often they can't even hear that they're singing badly — they know only because they've been told. Fortunately, they make up just 2.5 percent of the population. (Researchers determine such statistics by studying small groups of people, often in one country or even one city, but they generally apply the results to the global population. After all, we are all human no matter where we live.) Despite how uncommon amusia is, we frequently refer to bad singers as tone deaf. The term is an easy catch-all and sounds less complicated than motor control or timbre translation problems.

The good news is you probably aren't tone deaf. In fact, you're probably a better singer than you think you are. Yes, by Hutchins's measure, 60 percent of us are bad singers, at least some of the time. But this is a relatively new area of research and there's not yet a consensus on what exactly is good singing and what is bad singing.

Pfordresher, working with a more generous range of what's acceptable, has encouraging news for wannabe singers. In 2005, he was teaching at the San Antonio campus of the University of Texas. He and colleague Steven Brown asked more than one thousand first-year students if they could imitate a melody by singing it. A whopping 59 percent said they wouldn't be able to. Next, they asked eighty students—first in Texas, then at the University of Buffalo—to do it. After singing "Happy Birthday" and doing other warm-up exercises, the participants tried to match the melodies, which were just four notes long and within a small pitch range. Matching melodies is actually easier than matching individual notes, and only about 14 percent were consistently off by more than a semitone (the difference between adjacent black and white piano keys—say, A and B-flat—or between any two frets on a guitar).

Pfordresher and Brown had asked the question in the first place because they were looking for a way to find more bad singers to study. They eventually gave up. There was no reliable relationship between how people thought they'd do and how they ended up doing, so it wasn't much help. The good news: "People are often pleasantly surprised to find that they're not bad enough to be in our experiment." The bad: finding inaccurate singers remains a big problem for his lab.

As part of their research, Pfordresher, Hutchins, and their colleagues listen to a lot of poor-pitch singing. I once asked Hutchins: "Is that painful?" But he said it wasn't so hard, especially since he'd developed a good tolerance for it.

Besides, part of the reason he ended up doing this research was his appreciation for people trying to produce music, regardless of how good they are. "I genuinely like listening to people making music," he said. "And one of the side benefits is I can listen to anybody singing a song on the street and I can say, You know, I've heard worse."

Hutchins has also studied what he calls "the vocal generosity effect" and discovered that listeners considered vocals in tune when the singers were within half a semitone of the correct pitch, but were less willing to accept such variations from violinists. So that's a little scientific evidence to back up all those vocal coaches who tell us not to be so self-conscious.

Still, the inner voice that insists we're terrible at singing is a powerful force. But Lorri Neilsen Glenn, a former poet laureate of Halifax, believes we shouldn't stop doing things we love just because we aren't good at them. In fact, she encourages bad singing. "I've been singing enthusiastically off-key all my life," she told me. "I think of it as colouring outside the lines."

I believed Barnes when he told me I was trainable. I was already learning lots from him and not even I am so impatient that I expected to be able to sing after just two lessons, but I couldn't help being hard on myself. Perhaps a little reassurance that I wasn't actually tone deaf would goose my confidence. For that, I needed to go to the scientists.

The Science of Tone Deafness

"I've Got News for You"

Isabelle Peretz made me an espresso in her office.
This, I would soon realize, was the equivalent of giving a
cigarette to a man about to face the firing squad. Peretz is
a cognitive neuropsychologist and a professor at Université
de Montréal, where she is a founding co-director of the
influential International Laboratory for Brain, Music and
Sound Research, better known by the acronym BRAMS. It
was March 2011, and I'd travelled to her lab on the north
side of Mont Royal, the big hill that sits in the middle of
Montreal, to determine once and for all whether I was
really scientifically tone deaf.

I wasn't particularly nervous as I walked from the Metro
station up to the former convent that's now home to BRAMS
on what I, as a McGill grad, always considered the other
side of the mountain. True, I hadn't done all that well on
the lab's online test. I'd had to listen to pairs of short melodies and then indicate whether they were the same or

different. (I later learned that in half of the second melodies a single note was different.) If the tests had, say, played two sentences— "This, I would soon realize, was the equivalent of giving a cigarette to a man about to face the firing squad" and "This, I would soon realize, was the equivalent of giving a cigarette to the man about to face the firing squad"—I'd have no problem noticing that "a man" had become "the man." But when I was comparing melodies, I was often flummoxed. By the time the second melody started, I'd couldn't remember the first one well enough. I blamed it on my inability to focus. Since there are only two choices each time, a random result is 50; in other words, a completely deaf person would, on average, get 50 on the test. In the block I found the hardest, I managed to pull off a 54. This result brought my score on the three tests down to 69 percent. I knew that wasn't technically a pass, but I hoped it was close enough that Peretz might consider me just woefully untrained. A gentleman's C kinda thing.

The lab is in a beautiful building with hardwood floors and none of the institutional look and feel I expected from a modern university research centre. When I arrived on Monday morning, I met Mihaela Felezeu. Originally from Romania, she has a serene demeanour, which I appreciated under the circumstances. She took me into a large room, sat me down in front of one of the three computers, and gave me a set of headphones. The results were sometimes different in the lab, she explained, because there were fewer distractions than when people do them online. This was my chance to redeem myself. But I didn't and, despite a complete lack of evidence, I again blamed my wandering mind.

One of the tests that was new to me measured my ability to detect meter, the rhythmic structure of a composition. The task was to decide if a short segment of music was a waltz or a march, which have different time signatures. A waltz is in 3/4 time: a strong beat followed by two lighter ones, oom-pah-pah, oom-pah-pah. A march, on the other hand, is a four-beat pattern, in 4/4 time: a drill sergeant calling out "hup-two-three-four, hup-two-three-four."

I thought this would be easy, but the snippets all sounded like waltzes to me. For another test, I had to listen to five tones and indicate whether the fourth was the same as the others. This seemed a lot easier—at least when the fourth tone was quite different from the others. But I realized I probably wasn't doing so well when the fourth tone was only slightly different.

If the listening session wasn't embarrassing enough, Felezeu wanted me to sing. First, she recorded me butchering "Happy Birthday." Then, piling on the mortification, she wanted another rendition of the song, but with *la la la* instead of the lyrics. Finally, I had to sing *aaah* from low to high to low again. I sure was glad when it was over.

Fortunately, the next set of tests was a better fit with my skills. One was a spatial exam, which started off easy and then became tougher and tougher. Called the Matrix Test, it assesses visual perception and problem-solving ability by presenting unfinished geometric patterns and asking the subject to complete them with one of five possible solutions. Felezeu then recited a string of numbers, which I had to repeat backwards. The strings became longer and longer as we went, but I could tell by the look on her face that I was acing this one. At least, I think that's what her face was indicating.

After two hours of testing, Peretz invited me into her office and introduced me to researcher Nathalie Gosselin. The three of us chatted about singing and music as Peretz worked her magic with her espresso machine. Then she said, "So, the verdict…"

By this point, my early morning confidence was beginning to melt. "Oh, dear."

"No, I'm kidding. The coffee first…" And then she giggled, though not unkindly. She has a charming, gentle laugh.

"Like in Hitchcock," Gosselin said, "we create tension with music."

Of course, I didn't know Peretz shared my love of a good espresso when I first contacted her in 2011. I just knew she was a pioneer—the pioneer, really—in tone deafness research, though tone deaf is not a term she uses, preferring the one she coined: congenital amusia.

Born in 1956, Peretz grew up in Brussels, where she studied classical guitar for eleven years. Eventually she figured she'd be better at science than music, though she kept her guitar. A few years ago, she began taking singing lessons with the goal of joining a choir. She was motivated by the prevailing theory that singing with other people releases endorphins. She thought she needed lessons first because she wasn't confident in her abilities; her experience was limited to singing to her children when they were small.

When I saw her at the 2013 Society of Music Perception and Cognition conference, a biennial gathering of the world's leading music researchers, which just happened to be in Toronto, I invited her to join me for an espresso at a nearby café. She said the lessons were really making a difference to her mood. Several months later, though, she told me that after a year of lessons, she had decided to stick to the guitar. But because she believes that music shouldn't be played or listened to in isolation, that it's really something to do with other people, she joined a guitar orchestra. "I'm totally immersed and back to what I was doing when I was young," she said. "That's a lot of fun."

In 1985, armed with a Ph.D. in experimental psychology from Université libre de Bruxelles, she landed at Université de Montreal. In 2005, she created BRAMS with Robert Zatorre of McGill. Originally from Buenos Aires, he did a Ph.D. in experimental psychology at Brown University and then started a post-doc at McGill in 1981. He's been there ever since. Montreal already had a strong reputation for music research before BRAMS, and the new lab only increased the city's standing as a leading cluster.

Canada is at the forefront of the rapidly expanding field of music cognition with people such as Glenn Schellenberg, who runs the Music & Cognition Lab at the University of Toronto Mississauga; Sandra Trehub, who studies the development of listening skills in infants and young children at the same school; Laurel Trainor, who researches musical development in children and infants at McMaster University in Hamilton, Ontario; and Caroline Palmer, Canada research chair in the cognitive neuroscience of performance at McGill. Outside of Canada, leading researchers include Aniruddh Patel at Tufts University in Medford, Massachusetts, Carol Lynne Krumhansl at Cornell University in New York State, Stefan Koelsch at the University of Bergen in Norway, and Bill Thompson, who was at the University of Toronto until 2007 when he moved to Macquarie University in Sydney, Australia. Other top experts who specialize in amusia research include Gottfried Schlaug at Harvard and Lauren Stewart at Goldsmiths, University of London.

For a long time, many people thought tone deafness, as a neurological condition, was just a myth. But while studying patients who'd lost their ability to appreciate music after stroke or trauma had damaged their brains, Peretz began to wonder if it was possible that some people were born this way. Finding a case proved harder than she had imagined. Using newspaper ads to find subjects (since the Internet was not yet ubiquitous), Peretz tested many people over many years and began to doubt she'd ever discover one.

But then a French-speaking woman in her early forties, a former nurse working on a master's degree in mental health, answered an ad. She described herself as "musically impaired." Under social pressure, she'd been in a church choir as a child and, later, in a high school band. But after she married a college music teacher, she became acutely aware that she didn't like listening to music because it sounded like noise and stressed her out. Testing showed that the woman couldn't detect variations in pitch smaller than two semitones. Of the thirty-seven people Peretz had studied closely, here finally was the textbook example she'd been looking for: the first documented case of what she would call congenital amusia. In a 2002 paper on the woman, known as Monica in the study, Peretz and her colleagues stated, "From the data presented, we conclude that congenital amusia, or tone-deafness, is not a myth, but a genuine and specific learning disability for music."

At the time, the prevailing theory, especially in music education circles, was that everyone was musical and could learn to play an instrument. This jibed well with another popular idea at the time that anyone, with enough practice, could be a high-performing, high-achieving musician. Peretz's discovery threw a little scientific stink bomb into this egalitarian view. Though it a was shocking finding at the time, even for Peretz, it wasn't long before researchers all over the world were studying the disorder, or as researchers sometimes call it, the deficit.

Regular folks like me sometimes think scientists and academics get their jollies coming up with fancy terminology that only they can understand, but Peretz has good reason to avoid the colloquial name and not just because tone deaf is a dis in other fields, including politics, corporate public relations, and personal relationships. People call bad singers tone deaf even if their problem has nothing to do with an inability to hear pitch. Peretz talks to so many people who mistakenly think they're tone deaf that her guess is that educated overachievers — her fellow university professors, for example — are so accustomed to succeeding at everything that they think there must be something wrong with them if they are not naturally gifted singers.

Lots of people also tell me they are tone deaf. Though

I've sent many the link to the BRAMS online test, only one of them failed. Or perhaps more accurately, only one admitted to me that she'd failed. (The news didn't seem to bother her; she told me in a Facebook message: "It's okay if I am tone deaf. I gave up on a music career a long time ago.") Usually, I can squelch a self-misdiagnosis just by asking if the problem is simply that they can't sing well or if they also have trouble telling pitches apart. For many years, researchers believed that 4 percent of the population was amusic, but recently Peretz's lab did a broader study and found that, according to the criteria she uses, it's even rarer: just 2.5 percent.

Although both "tone deaf" and "tin ear" — another common term for pitch impairment as well as for mocking politicians and others — suggest the trouble is aural, it's actually in the brain. Peretz also liked the term congenital amusia because it was broad enough to cover pitch perception difficulties, rhythm perception troubles, and other music processing problems. Though it's much rarer than pitch deafness, some people are beat deaf and have trouble with rhythm. (But just because someone is a terrible dancer, doesn't mean he's beat deaf, just as not all bad singers are tone deaf.)

Peretz's term builds on a much older one. In 1888, German doctor and anatomist August Knoblauch came up with the first cognitive model of music. It wasn't entirely accurate: he thought music was a left-hemisphere function, for instance, and we now know that the networks that support it are distributed across both hemispheres. But some of his ideas were prescient. He split music disorders into sensory impairments (perception) and motor ones (production), and used amusia to refer to the latter when caused by lesions to the motor centre. He proposed the existence of nine forms of the condition, including "note blindness" (problems reading musical notation) and "note deafness" (trouble comprehending music). Knoblauch's paper spurred a lot more research into music and the brain in the following years, and amusia came to mean an impairment in music abilities due to brain damage. Today, acquired amusia is the kind you get after a stroke or brain injury; congenital amusia is the kind you're born with. It's similar to disorders such as dyslexia and other learning disabilities.

While this area of research seems a strange one for someone who might have become a musician, Peretz doesn't see it that way. "It allows us to better understand the mechanics behind pitch perception and music processing at the behaviour level, at the cognitive or theoretical level, at the computation level, at the brain level, and at the genetic level," she said. "What else can you hope for when you are in psychology?"

I'd had two lessons with Barnes and they had con-firmed in excruciating fashion what I already I knew: I was a dreadful singer. The recordings of our sessions made for painful listening, even for me. But the most common form of amusia isn't just about bad singing. The disorder affects the perception, memory, and production of pitch—in other words, hearing, remembering, and singing. Though speech also relies on changes in pitch, amusia probably affects only music processing (but admittedly, there isn't consensus on this). Sufferers can't tell the difference between pitches that are close together, what Peretz calls "the discrimination of fine-grain pitch differences." While these small changes may be crucial in music, they aren't so meaningful in speech. The frets on a guitar are a semitone apart. In speech, pitch changes are usually six or seven semitones or more. When you ask a question, for example, the last word or syllable typically goes up as much as an octave, or twelve semitones.

The listening tests I'd completed online were part of the Montreal Battery of Evaluation of Amusia. Based on tests Peretz had used to study people with brain damage, the battery is now used by researchers around the world. The six blocks measure the ability to perceive contour, interval, scale, rhythm, meter, and memory. I did all of them in the lab. For most people, the tests are easy: even someone without any musical training will typically score in the 90s. So an average score in the low 70s, while technically a pass, indicates problems.

Then there's the singing. While many amusics may be unaware of their inability to hear pitch and remember melody, few of the afflicted don't know they're bad singers. The diagnostic question is usually, "Do you sing in tune?" And the answer is often, "I don't know, but I was told I don't." When researchers ask subjects to sing a well-known song, often "Happy Birthday," they can do it, though out of tune. But when they have to replace the words with *la la la*, most are lost after the first two notes. Without the lyrics, they just can't find the song's contour — the shape of the melody—because they don't have a representation of it in memory. (Weirdly, though, there are a few amusics who do sing in tune. They just aren't aware of it.)

Still, three musicians, including a professional singing coach, had listened to me try to match pitches and concluded that I was trainable. Plus, I knew the stats: although 17 percent of people believe they're tone deaf, most of them aren't. So, despite my weak performance in the lab that morning, I still expected to hear a "you just need training" diagnosis from Peretz.

Finally, she handed me an espresso. Then she gave me the news I didn't want to hear: I was, the tests made clear, a typical amusic—part of a tiny, hopeless minority.

This is how we hear music: sound waves go in the outer ear and pass through the ear canal until they reach the eardrum, causing it to vibrate. These vibrations move tiny bones, called ossicles, in the middle ear, transferring the movement to the cochlea, a cavity shaped a bit like a snail, located in the inner ear. The wider base of the tapered spiral responds to higher frequencies, while the narrow end picks up on lower ones. Fluid inside the cochlea vibrates tiny hairs, called cilia, which stimulate thousands of hair cells with auditory nerve receptors in the organ of Corti, sending signals to the brain, which interprets the signals as sound.

Presented like that, hearing seems incredibly complex and difficult. Most people never have to think about it, just as we never have to think about how our eyes or our knees work. And that's a good thing. For a few of us, though, something goes wrong.

Of course, the hitch for amusics is in the brain, not the ear. We don't process what we hear as well as we should. Impaired pitch recognition and production suggests a problem in the brain's action-perception network or, more specifically, a neural structure called the arcuate fasciculus. When researchers at Beth Israel Deaconess Medical Center and Harvard Medical School compared brain images of amusics and regular folk, they discovered a difference in these pathways.

So I have a bad connection in my head. And unfortunately, I can't just hang up and dial again.

The tests showed that I had the most trouble with the melodic organization of music. I later learned that I'd done poorly on contour, interval, and meter (the Waltz or March test I thought would be easy, but quickly discovered wasn't). And I was just below a pass on scale (major or minor mode). But I'd crushed the rhythm test and, much to my surprise, did okay on the memory test. The latter required me to listen to several melodies and indicate whether or not I'd heard them earlier in a battery of tests. Except for that test and the metric one, all the blocks required comparing two melodies. As Peretz explained, if my problem was limited to working memory, required for holding a melody in memory for comparison, then I should have failed the rhythmic test. So my weakness is related to retaining musical pitch information in my head.

I like to think I have a knack for recognizing songs and would have been a strong contestant on that old TV game show *Name That Tune,* but holding pitch information in memory would soon become a problem in my singing lessons. For the longest time "Amazing Grace" continued to stump me, despite singing it many times with Barnes and practising it on my own. I could recognize the song immediately if he played it on the piano, but if he asked me to sing it, I had no idea what the notes were supposed to be until I had committed to learning the contour.

Meanwhile, in the five-tone test, I was able to detect a quarter-semitone change in pitch less than 40 percent of the time. So while not all notes sound the same to me, I can't tell the difference between two notes that are close together or if a singer is only a little off. But, to use an extreme example, if I go to YouTube and watch the execrable rendition of "The Star-Spangled Banner" Roseanne Barr performed before a San Diego Padres game in 1990, I can tell she wasn't singing the right notes and doesn't have the melody right at all.

My average score on the Montreal Battery was 70 percent, which put me two standard deviations away from the mean of the whole population. A term from statistics, two standard deviations indicates how far outside the normal range I am. But the disorder isn't simply there or not there; it's a continuum. Some people have trouble with pitch and some people have even more trouble with pitch. Peretz showed me where I was on a graph: I was close to the edge of the line between amusics and normal people, but definitely on the wrong side of it.

"So this means what?" I asked her. A stupid question, I guess, but also a reflexive one.

"We would love to test you more," she said. "What it means for your singing training, we'll see."

That afternoon, I joined Sean Hutchins for more testing. He'd originally come to Canada to do his Ph.D. at McGill and, when I first met him, he was doing a post-doc at the BRAMS lab. He took me into a small room with a computer, headphones, and a short Plexiglas plank covered in green tape with a beige strip down the middle. One of the questions Peretz and Gosselin had asked me was "Why singing?" I'd have a less challenging go of it if I took up an instrument—though if the one that Hutchins and his colleagues invented is any indication, less challenging is a relative term. They call it a slider. Hutchins told me, "It's designed to be the easiest instrument in the world."

The slider allowed him to study why some people are bad singers by separating their ability to match pitch from their ability to control their voice. If you master the instrument, pitch perception isn't your problem, but maybe motor control is. To play the slider, you press the touch-sensitive strip, which is a position sensor overlaid on a pressure sensor. Designed to work the way the human voice does, the strip generates a continuously changing pitch with a range of one octave, from low on the left end to high on the right end. For the experiment, Hutchins asked me to put on headphones, listen to a pitch provided by the computer, and move my finger along the strip until I found the spot where the slider's note matched the computer's target note. At that point I was to hit the space bar on the computer to indicate I'd found it. Then the computer would generate another target note and I would try to match that one.

Although he would measure how long it took each time, he was mostly interested in how precise I was.

When he explained all this to me, I said, "Okay, this will be fun." I wanted to do well and I wanted to prove that despite the confidence-shattering diagnosis Peretz had just given me, I really was musical, I really was normal.

For a musician or anyone with a good ear, this is indeed an easy instrument to play. But I found it a struggle. The trick was that I wasn't allowed to listen to the target note and the slider's note at the same time, so I had to keep lifting my finger off the slider to check the target pitch again. I found some tones harder to match than others. Sometimes I would start off fairly close to the target; other times, I'd begin nowhere near it, at the completely wrong end of the strip. And a few efforts took what seemed like forever to match the pitch — and even then, occasionally I was just guessing.

Next, I had to sing. Hutchins recorded me singing *baa* at five different pitches: high, medium-high, medium, medium-low, and low. Not specific notes, just five tones that spanned my vocal range. I held each note for two seconds, trying to keep the pitch steady. As I did this exercise, I wasn't always entirely sure if my notes were high or low or whatever. Fortunately, he figured out a way to use what he'd recorded. I put the headphones back on so I could listen to my *baas* played back to me randomly and try to match each one as closely as possible. Since someone else's voice has a different timbre than yours, matching it requires not just figuring out what the note is but also translating it to your own vocal timbre. Matching your own voice is simpler. There's no timbre translation to worry about, and there's no question whether you can hit the notes.

Although I didn't need a scientist to tell me I was bad at this, I asked Hutchins how I did. He said he'd have to go over the data to be sure, but his preliminary analysis suggested I hadn't done too badly on matching pitches with the slider. "There were a few errors, but it was generally pretty good," he said. Then he added, with generous understatement, "But on matching yourself, there were quite a few more errors."

Desperate for good news, I found this encouraging. If I was hearing better than I was singing, maybe it was more of a motor control problem. Not so fast: it didn't rule out a perceptual problem, he explained. And it didn't rule out what the ultimate cause of the motor problem was — it may have developed because of some low-level perceptual problems. "But it would say that there is some problem in producing consistent pitch."

Johann Sebastian Bach wrote some or all of his famous cello suites while working for Prince Leopold's small court in the German city of Köthen. Bach was Kapellmeister — the person in charge of music making — from 1717 to 1723. And for a time, the palace boasted a good, well-funded orchestra. But soon after Leopold married his cousin Friederica Henrietta in 1721, his love for and financial commitment to music cooled. This may have been simply because the nineteen-year-old princess wanted more attention and was jealous of the time and money the prince spent on music and with his musical employee, according to *The Cello Suites: J. S. Bach, Pablo Casals, and the Search for a Baroque Masterpiece* by Eric Siblin. But writing to a friend, a bitter Bach called the princess an "amusa," someone with no interest in music. After I read that, I emailed Peretz to see if she'd ever heard the term; she hadn't, but noted, "Bach had vision! To say the least."

Amusia, following the neurological nomenclature, means "without music." And, indeed, a lot of amusics are unmoved by music. A few even find it extremely unpleasant because it sounds like a cracking noise and they do their best to avoid it. But most are indifferent. They don't want to listen to music, and going to a concert is like listening to a speech in a foreign language. Few like to admit this publicly, though, for fear other people will consider them inhuman. In the lab, BRAMS researchers have heard subjects talk about their indifference and then, an hour later, say on camera, "Of course I love music!" One amusic told Gosselin that, during his divorce, his wife said, "You don't under-stand emotion. The proof of that is you don't like music."

But the two enduring passions in my life are hockey and music. Peretz and Gosselin were surprised to hear this. Skeptical, in fact. They asked if I liked only certain kinds of music and suggested I might merely be responding to the words. But I was convinced that made no sense: I was clearly hearing more than the words. I pointed out that if, say, the opening motif of Beethoven's Symphony No. 5 or Deep Purple's "Smoke on the Water" — to cite two really obvious examples — started playing, I would recognize them immediately. Peretz suggested that I may be picking up on other cues. The tests showed that my rhythm perception was fine, for instance. She asked me if I danced. I said that when I'd done the online test and came to "Would you consider yourself to be a good dancer?" in the questionnaire, I asked my wife. "For a guy," she said, "you're a very good dancer." So I put down moderate.

The more I talked about the role of music in my life, the more intrigued Peretz became. "I am stunned. And I've seen many amusic cases," she said. "I would love to be in your brain."

The inside of my brain was not a serene place in the days after my diagnosis. It's one thing to be afflicted with something common; it's quite another to learn that you're some kind of subhuman freak. I was devastated. This was science, after all. It felt as though someone took a measuring tape and told me I was so many feet and inches tall. And while I am not a scientist—I flunked out of mining engineering after two years—I am definitely not the kind of person who doubts science.

Later, my psychologist friend Alex Russell suggested I shouldn't take the news as a life sentence, because diagnoses can shift and change over the years. "I would not want to dampen your enthusiasm or even your confidence about being able to be a music maker," he assured me. But for a long time, I was reluctant to tell anyone I was amusic. I like to think I have a reputation as a knowledgeable music fan—someone who turns friends on to great new albums, puts together much-appreciated mix CDs, and buys the tickets and rounds people up for shows. What would my friends think when they learned I really was tone deaf? I'd be ruined. Even I started to wonder, Am I just a poseur?

One of the first people I told was my friend Timm Hughes, an actor who describes himself as a two-and-a-half threat: acting, singing, and in a pinch, dancing. It was several months after my diagnosis on a cool, grey November day at another friend's cottage. We'd gone for a slow walk on the granite outcrops, because Hughes was recovering from a hernia operation. After he talked about his condition, I decided to open up about mine. His reaction: "I always wondered why you liked The Jesus and Mary Chain." That was exactly the kind of response I feared.

For the first week after I returned from Montreal, I didn't practise at all. Although I knew I couldn't sing before I'd gone there, the diagnosis was like saying: We checked your brain and you'll never be able to sing. With a week to go before my next lesson, I did put in a few hours, working along to the recording of the previous session. After I showed up, Barnes spent most of the time trying to talk me out of my funk. He knew that an emotional setback was a real setback, especially in singing, where so many of us have baggage from when we were young. Barnes often used his experience at the gym as an analogy for me. When he first started working with a personal trainer, he had to confront his own emotional demons: falling off the vaulting horse in grade seven gym class because he wasn't co-ordinated, for example. "When I go to a gym now, I'm clean of all that shit. It's gone. I know what I'm doing, I've emotionally rescued myself from that hell."

He'd tested me the first time we met and decided I was trainable, and no scientists were going to convince him otherwise. "You've been told you're one of those guys who can't hear notes properly. Nothing new for me. New for you?" he said. "But all this is completely moot because they're not studying whether you're trainable." So the diagnosis didn't change his hypothesis at all; it just reinforced what we already knew: I'd have to work hard. "The worst thing about all this is the emotional, underlying shit that stops us from learning. That's all I ever deal with, whether it's you or a recording artist or a Broadway actor," he said. "What I'm watching for is whether you get downhearted, whether you go away saying, This fucking sucks and I hate that guy and I can't do this. The fact that you love music is a compelling argument for your ability to focus on this."

During the first two lessons, we'd worked on breathing and other exercises, and I hadn't sung any songs. It felt like conjugating verbs while learning a new language, instead of having a conversation. Today, we sang. We batted a few song titles around, some of which Barnes said were too hard. And then we settled on "Before the Next Teardrop Falls," a song I've loved since Freddy Fender had a big hit with it in the 1970s.

That made the lesson more fun, but I couldn't shake the question that had nagged me since my diagnosis: *Am I like the colour-blind guy who wants to be an interior designer?*

"Tell It Like It Is"

And so I was at war with my voice.

In my last lesson, after I'd sung along to "Before the Next Teardrop Falls," Barnes had asked how I thought it sounded.

"It sounds the way it always does when I sing along: great," I replied.

But when I heard the recordings of my lessons, I knew I didn't sound anywhere near great. My practice regimen, as haphazard as it was, consisted of singing along to the previous session. And that meant hearing it again and again and again. When I showed up the next time, Barnes asked how it had gone. I admitted that I knew I'd be off-pitch, but I was also hearing how far off the rhythm I was. "I heard some pretty bad stuff when I was practising the Freddy Fender song."

"You're not a good singer yet," he laughed, though without any malice, "so I'm not surprised. But did you hear it when you got the note?"

"I can tell when it's way off, but when it's close, that's harder for me."

We worked on "Before the Next Teardrop Falls" again, trying to get the rhythm by talking the lyrics, not worrying about pitch. I didn't get the notes right away, but I found them eventually, which kept Barnes convinced that I was trainable. He told me this often and I think he believed it, but he also knew he had to convince me that it was possible for me to sing. Without the hope of possibility, it wouldn't be fun; in fact, it would be a miserable experience. After all, nobody enjoys doing something he's bad at. "Your language betrays you because every time I say you're getting the third one, you say, Yeah, it sounds horrible," he said. "Clearly, Montreal sent the very strong message that you can't do this."

After a while, he decided Freddy Fender was too high for me and we started working on "A Good Year for the Roses," a song made famous by George Jones. I knew it first from Elvis Costello, who recorded it on his album of country covers, *Almost Blue*. After the first run-through, Barnes noted that I knew this song a lot better than the Fender one — probably because I used to listen to that album obsessively. In the mid-1980s, a few years after *Almost Blue* came out, I was sharing a place with a friend. The neighbour upstairs was Kenny MacLean, the now-late bassist for Platinum Blonde, a successful Toronto pop metal hair band. Pubescent girls regularly came to the door of the duplex to see if MacLean was home, and I felt it was my duty to lie, telling them they had the wrong address. One day he said to my roommate, "Wow. Tim really likes that album, eh?"

Although she said it didn't sound like he was complaining about the noise, or even my musical taste, I felt bad that he heard me play it so often. But I did have to play it loud, so no one would hear me crooning along.

Always quick to offer positive reinforcement, Barnes regularly insisted that my ability to match his note was improving, that I was getting it on the second try instead of the third or fourth. Near the end of the lesson, Barnes told me, "Your ears are getting good."

"Getting better," I agreed uncertainly.

"Yeah. You can feel it? That's all that I ask. I don't care that you're there. If you can see improvement, then we are away to the races because it's going to take a lot of commitment and energy and focus. Your brain is being retrained, for chrissake."

A few days later, I was driving by myself to a friend's cottage in Georgian Bay. My iPod was in one of its ebullient moods, pumping gem after singable gem into my car's sound system: the Lowest of the Low's "Salesmen, Cheats and Liars," "Back Stabbers" by the O'Jays, and Neko Case's cover of Neil Young's "Dreamin' Man." As I bombed along the empty highway, I belted out each song loudly, gleefully—and, I soon realized, tunelessly.

"Shit," I said to the Canadian Shield. "The only thing these lessons are doing is making me hear what a terrible singer I am."

Later, I adopted a more optimistic take. If I can now tell that I suck, maybe Barnes is right and my ear is ever so slowly getting better.

I breathe all wrong.

"And that's why you sing out of tune," Barnes insisted. "If we break the code on this, singing will be easy." He had me blowing air through my lips so they made a sound like a motorboat or a little kid adding audio effects while playing with toy cars. *Bblblblblblblbl.* And I had to do it on his pitch. His idea was that this would ease some of the tension at work when I produce sound, tension created because I'm afraid I will sing the wrong note. Which, inevitably, becomes a self-fulfilling fear. And fear, along with improper breathing, was something I was going to have to conquer if I was ever going to learn to sing.

I tried to breathe from my diaphragm, make the weird sound, and match his notes, but it was too much for me. So he decided we'd use the breathing for relaxation exercises, not pitch-matching. In this case, he told me his teaching strategy. But sometimes when he would change exercises on me, I wondered if it was because I was doing something well enough to move on or because he could see it wasn't working and decided to try something new. I'm always intrigued by how other people teach, and I asked him about this. He admitted that he didn't want to overwhelm me or discourage me, but he also needed to push me so I'd start to feel I was improving. "It's an emotional reaction you're having to your voice, anyway," he said. "We don't want to stoke that fire, we want to stoke the other one."

He claimed to enjoy the challenge of coaching someone with so far to go and told me about working with a theatre group when he was younger. They were great actors but not-so-great singers, including a few who just couldn't

seem to sing on pitch. He was just twenty-one and a little
impatient. "I noted that with love and encouragement you
get there. A little bit of faith and a few simple rules and
even the worst singers can become part of a team," he told
me. "I got a lesson from the universe." A lesson that was
valuable when working with me. "This is not a test of my
patience so much as I'm learning another skill set. So that's
what keeps me on my toes."

We kept going, and I sang "A Good Year for the Roses"
for him. He was happy that my pitch was better—I was
hitting most of the low notes—but I was missing the high
ones in the chorus. He noticed that I wasn't physically getting air in between phrases. That was preventing me from
hitting the high notes. I admitted that I didn't know when
to breathe. "Whenever you're not singing," he said, "that's
when you breathe."

Instead of breathing properly, my hopeless attempts to
hit the high notes involved too much reaching with my
body, including straining my neck upwards, so he decided
to teach me a new singing position: the monkey. "You will
think I'm nuts," he said, "but I actually give it to all kinds
of professional singers." This position required me to bend
over, bend my knees, and dangle my arms. I did it and sang
"Amazing Grace." I felt ridiculous. We sang it together, then
he asked me to sing it on my own. I started to giggle and
said, "I'm just trying to hold it together here."

"You're doing a good job," he said. "You haven't shit your
pants, you're still wearing your clothes..."

Later, after I sang "A Good Year for the Roses," we talked
about pitch. Since I can't hear it well, I'm naturally always

curious about how other people hear. "For you," I asked, "is it just instinct that you know that's a higher note or a lower note?"

"I had to learn to hit the notes. Was my ear better than yours when I started—"

"Obviously," I interrupted. "But with this song, you're saying, 'You're okay on the verse, but too low on the chorus.' But how do I know that it should be higher? That's the ear part. I wish it was just, Okay, you have to learn this formula."

He assured me that if I focused on the notes, I would gradually improve my ability to hit them. Once again, he told me he was confident that I was teachable. "Do I think you can improve? Yes. Because you have improved. I believe that you can do this."

I wrote him a cheque for four more lessons.

Before my next session, my wife, Carmen, and I rented a cottage for ten days and invited our friend Kelly Crowe up for a visit. A CBC TV reporter, she grew up in small-town Ontario and as a ten-year-old she joined the United Church junior choir, which was short of members. But the choir leader asked her to stand in the back and insisted she not sing, just mouth the words. "And so that's what I did," she told me. "I've been mouthing the words ever since, too terrified to sing a note."

But Crowe did learn to play guitar. She brought it to the cottage and, because she wasn't going to sing, she insisted I do it. We did "A Good Year for the Roses" and Crowe was complimentary—she's a polite guest—but later Carmen told me the truth: I was off-key, but my voice was getting better.

Singing with Crowe made me realize that I never know when to come in when accompanying someone playing the piano or a guitar. I brought this up with Barnes: "When do I start singing?" I think he was trying to make me feel better when he said a lot of singers have that problem, so musicians will say, "I'll give you four bars and then you come in." But as much as that may help those with musical talent, it means nothing if you can't hear well enough to know when the pianist has played four bars.

After working on breathing exercises, we turned to ear training. I tried to match his piano notes with *mmm* as he moved up the scale. He wanted me to feel the vibration on my lips, because that indicated that I was breathing correctly and using my skull as a resonance chamber.

At the next session, a few lines into "A Good Year for the Roses," he stopped me. "Why am I starting you over?" he said. "You're not really breathing. Not really. Now I'm going to be tough."

I tried again and when I finished, he was happier. "That is good news," he said. "There was like two and a half lines that you didn't get, but you had the pitch on every other line." I was missing the highest notes, but I was also missing my phrasing pretty well everywhere. So he had me speak the lyrics to get the rhythm right. That went well, but he said I was just following Costello so we should see if I could do it with the piano. "Will you nod me in?" I asked. He did and we worked on the tune, note by note. And then I sang the whole song through and it went better than I had reason to hope it would. I wasn't pitch perfect, of course, but I made it all the way to the end without falling apart.

"That's it," he said.

"Okay, but I don't know if I could do it again."

"You probably can't. Right now... Here we go..." We sang it again, but I had run out of beginner's luck. Or maybe I was thinking about it too much because I wanted to do as well as I had the first time.

"Okay, lots of wrong notes," he said. (When I listened to the recording later, I realized with embarrassment that a lot of notes didn't just sound wrong, they sounded incredibly sour.) "But your phrasing's good and I can tell you know where they are supposed to be. So let's nail them."

Actually, I wasn't so sure I knew where the notes were supposed to be. But we did it again and he said, "We're getting your ears trained, brother."

At the end of the session, I asked him if I was at the stage where I could sing along with a friend. I explained that I was going to a cottage for an annual boys' weekend full of beer and guitars (and one harmonica). "Ummmm," he said, stretching it out. "Okay, why not. But it better be a good friend."

I went. I sang, though not loudly. No one complained. Beer and old friendships smooth out a lot of sour notes.

But the next time I visited Barnes, I whined about how bad I sounded on the recordings of our sessions. And about my slow progress. "I guess I'm still hoping for it to all of a sudden click in and *Oh, I can hear.*"

No miracle drug exists, of course, and this discussion

was another chance for Barnes to remind me that I was never going be an overnight success. "There's still a part of us that thinks, Add water and poof, I get more out of it," he said. "But you will have to continue to work quite hard in order to get to what I believe is your goal, which is to sing in tune, live, coming in at the right spot and knowing what you're doing. That's going to take time."

"But you honestly believe that's an attainable goal, with work and time?"

"I honestly believe it. If I was just taking your money…" He didn't finish his sentence. "We have proven beyond a shadow of a doubt that you can improve. The rate of improvement might be something that you're not happy with, but I'm relatively happy with it because every week I see it."

And he was hearing the improvement even though, he pointed out, I was not a great student. I didn't work hard enough between classes, and every class I would come in and say I wasn't getting better. "There's a crazy, courageous risk-taker inside of you that we have to wake up. We have to say to him, Yeah, you sing like shit. Now, will you sing a little louder? That dude has to show up."

He let me try singing Son Volt's "Windfall." While it's another favourite of mine, that made no difference: I was way off pitch and Barnes quickly switched to "Amazing Grace." Later, while we worked on "A Good Year for the Roses," he implored, "Don't be so damn scared."

Usually, about forty-five minutes into a lesson, Barnes would notice that I was losing my focus. And my brain did often feel like mush by that point. I had to concentrate so hard on listening to the pitches and, to be honest, sustained focus isn't a personal strength in anything I do. As a songwriter, Barnes could relate: "We get to walk away and then come back and no one knows we lost focus." Not so for singers.

On my next visit, I was feeling better about it all. Carmen, who is never shy about telling me what she really thinks, told me I'd improved as she listened to me practise. It wasn't long ago that she was laughing at me. Barnes started me off with some panting exercises and he was happy with my progress. "We know that you have made a major leap."

Later, he returned to the subject of fearlessness. "How's your courage level in the rest of your life?"

Average, I figured. Some things scare me more than others. Roller coasters terrify me, but I'm not afraid to ride my bike on busy downtown streets (even though plenty of my friends tell me I should be). And I'm not into extreme sports or other daredevilry, but every summer I go on a canoe trip, usually one that features fairly aggressive white-water. A class III rapid is scary, but the terror is half the thrill. A few years ago, my friend Alex Hutchinson wrote a piece about our trip on Quebec's Du Lièvre River for the *New York Times*. The article ends with my canoemate Steve Watt and me attempting a set of rapids we probably shouldn't have tried and dumping in spectacular fashion. (It felt like our canoe had ejector seats.) After my mom read the story, she said to Carmen, "Do you think these trips are too dangerous for Tim to be going on?"

I mentioned this to Barnes, adding, "This is scarier than

going down a river because this is—"

He finished my sentence for me: "This is about you."

For all his understanding and encouragement, though, I just couldn't seem to build my confidence. Barnes might have been able to track my improvement, but I couldn't. I told him, "I can't say, 'I'm still a bad singer but I'm getting better.'"

"But you know what? You've held that perspective from day one," he said. "You haven't been able to hear yourself getting better at all. I challenge you that that's something you hold in your head about your singing. And you've often suspected that I'm lying to you about you getting better and you can't tell." It was a familiar refrain by that point, and we both knew it was going to be a long slog.

He had me lie down on the couch and make *ha* sounds on his note, but I wasn't doing it right.

"So listen once. *Ha ha.*"

I screwed it up again.

"Listen for a minute, just listen. You're making ugly noises that aren't really the sound."

I tried again.

"What's the difference between what I'm doing and what you're doing?"

"Yours sounds good."

"Yeah. How? Why?"

"Breathing."

"So breathe," he said.

Faced with what I was going through, I imagine most people would just give up and get their art fix somewhere else: painting or photography or poetry. But when you love something that much, you want to create it, not just consume it. I love watching hockey — and talking about it and reading about it — but nothing beats playing it. I know that someday, unless I die suddenly on the ice, I will have to give up playing. Many of my friends have already quit due to bad knees or hips or shoulders, but I know others who keep playing even after hip and knee replacements. They aren't ready to give up the game they love, and I totally get that. I dread the day when I can't play anymore.

And so, here I was trying to make music when any sensible person would say there are lots of other hobbies that I might be good at. But that's the point: music isn't just a hobby to me.

Good thing, since my vocal lessons weren't exactly fun. I didn't dread them, though. Barnes continued to be patient and encouraging, and we made each other laugh a lot. I always felt better when I walked down the stairs and out onto Queen Street after a session. It wasn't just relief, though I'm sure that was part of it. I felt I'd accomplished something.

Practising, however, was another matter. I resorted to the methods I'd employed as an uninspired high school student: I crammed. Right after a lesson, I'd think, *Oh, I survived that so I don't have to do anything.* I ignored my exercises until a few days before the next lesson, then went hard. Not hard enough, apparently. Barnes had talked to me about this before. He wanted me to practise every day, even if it was for only ten minutes.

"I know, I know," I said.

"No, no. No more 'I know, I know.' You've got to do it. If you were somebody who was a really good singer who was trying to get better, you'd still have to practise consistently. You're a bad singer. You need ten minutes a day. Take it from me, it will not get better. You can't cram. It doesn't work."

By my eleventh lesson, Barnes began to show some previously contained frustration. I still wasn't breathing from down in my diaphragm. He sent me to the couch to lie down; he asked me to create a balloon of air with my diaphragm and then press my fingers against it. But I still found trying to breathe properly at the same time as I was trying to match his note too difficult. Doing just one was hard enough, but both at once was beyond me. He told me to stand up so he could put his fist against the balloon as I made panting sounds on my own note again and again.

"Work against my fist," he said. "Drop your head. Can you feel how tense you are?" I could. "Your body tension is keeping you from singing, my friend. Your entire body is tense. Your neck position is tension. You're breathing up here," he said, pointing to my upper lungs. "You're, like, freaked out. Physically freaked out."

"Okay," I said.

"That's the number one thing you've gotta change. That's all you have to change."

In an effort to get the tension out of my body, he asked me to do the monkey. And then he did it with me. It took a while and a lot of cajoling on his part, but I finally started doing it well enough to satisfy him. "Yup, that's it. Keep going. I'll start playing notes for you in a minute."

"Are you serious?"

"I am. I'm finally seeing your shoulders come down and I'm seeing your head finally relaxing."

When he said he wanted to show me a new position called the fetus, I laughed. Which probably wasn't the right thing to do under the circumstances. That's when he started to drop the nurturing approach. "You have to get serious about this," he said. "You're not really engaged with this process. You haven't jumped in. It's a heavy thing for you to go from 'I can't sing' to 'I can sing.'"

For the fetus position, I had to get down on my haunches, a bit like child's pose in yoga, except I had to put my elbows between my knees and hold my cheeks to my palms on Barnes's red rug. Doing this was trickier than it sounds. Then I had to pant.

We panted back and forth until he pointed out that I was starting on his note but going sour as I tried to hold it. "Can you hear it?"

"I wasn't really listening," I admitted. "I was just thinking my face hurts."

"You're very complainy for someone who doesn't practise," he said. This exercise would really help release tension, he pointed out. "So it's going to be your homework. Get used to this position. I have to be a little harder with you. You're too lax on yourself; you're not going to get there. It's going to be hard work."

I did a couple more positions for him, including one

called the hangover, which involved getting on my hands and knees and letting my head dangle. He told me he'd put himself in the hands of coaches who made him do crazier things than this and were a lot harder on him than he was on me. "I've been going easy on you, and I can't do that anymore because it's not working," he said. "You're not taking it on. You're still dancing around it and don't think you can do it. I'm not seeing you go, Okay! I'm seeing you go, That's ridiculous. When I say, how was that? You essentially go through the same trip. You go, *I'm never going to get this, this is ridiculous, I don't have what it takes.* What that tells me is we have a mountain of stuff you have in your heart and head about this before we ever get to the work."

He didn't yell or sputter or go red in the face — he seemed frustrated and disappointed, not angry — but he did go on. "You're not trusting the method. You're second guessing the whole thing. Trust me, go away and do the homework or quit and do us a favour. What do you think I do this for? Just to laugh at you?"

"It's a bit of a bonus, I guess."

He ignored my glib reply. "I can't coach you until you jump in the pool. Come back and show me: *I'm fucking scared and I hate this but look what I did.* I want to see that fire from you because we're getting somewhere but we're never going to get there. I may not have the skill set to get around your defences on this. You believe this is hard and you can't do it. You're not racking up your wins the way I need you to."

This was Barnes playing the hardass. As chewing-outs go, it was a lot tamer than I'm used to. And yet, he apologized the next week. He shouldn't have — he was right about me not working hard enough. I'd been taking advantage of his patience and his willingness to take on a student who really had no business studying with him.

"Can You Hear the Music"

A few years ago, as I faced a urinal, minding my own business, a guy with a soul patch a few pieces of porcelain down from me turned and asked: "Aren't you a little old to be here?"

Here was the Kool Haus, a cavernous venue near Toronto's waterfront, and perhaps I did look too senior (though we were both there to see Nick Cave, who is older than I am). I regularly go to club gigs where I am, alas, old enough to be the father of everyone else in the crowd. I'm not that interested in — or have already seen, usually in a more intimate setting — most of the acts that play big venues. So my idea of a good time is to stand, cold beer in hand, amid the crush of the crowd, bopping and quietly singing along to great music.

I do know a few other middle-aged teenagers — and not all of them are aging rock critics — but most people my age roll their eyes when I mention what I listen to.

Meanwhile, I'm convinced the young concertgoers around me are thinking, Who brought her dad? Of course, if I ever wanted to push my way to the front, I could always just announce in a slightly panicked voice, "Where's my daughter? I need to find my daughter."

Aside from the soul patch kid, most people are polite enough not to say anything. But when I gave the stare to two women behind me who were yakking their way through a Calexico show at Toronto's Mod Club, one said, "Come on, let's move away from this middle-aged party." Whatever, I figured. One thing about being old is I actually try to enjoy the music. I mean, who goes to Calexico to be seen? A friend says the age difference stops bugging her as soon as the lights go out, but I'm more self-conscious. I cringe when someone bumps into me and says, "Sorry, sir."

I asked the guy at the nearby urinal, "How old is too old?" That was a question I really didn't want the answer to. Unless I am at an overpriced restaurant, going out in Toronto means wondering what happened to the rest of my generation. Nowadays my friend Bill Reynolds (a former music critic) and I go to shows together and laugh at the overparented kids who wear earplugs. Shouldn't they act as if they'll never grow old — or at least die before they do — while we take care of ourselves because we totally get what Leonard Cohen was on about when he sang that he ached in the places where he used to play? All I can say is: if I go deaf at eighty, it was all worth it.

My favourite place to see a band is the Horseshoe Tavern, which is like many music clubs in cities all over North America: cramped, sweaty, bad washrooms. Other venues in my hometown have clearer sightlines and cleaner sound, but none can match its nearly seventy-year history of hosting bands. No one bristles when the place calls itself the Legendary Horseshoe Tavern.

Sometimes, especially when I move up close to the stage, I stand there in bliss, but also in wonder: the music might as well be magic to me. I don't mean that in a trite *that song is magical* way. I mean that when I watch people make music, it's like an advanced technology beyond my primitive comprehension. I'm perplexed about how they do it. I can watch great hockey players and marvel at the speed or the moves or the shots, but because I can skate—and score the occasional goal—that doesn't seem like magic. It's just excellence I know I'll never match.

Music is different. The drummer, I get. But when I stare at the guitar player's hands—one strumming, the other holding strings against the fretboard—I can't make the connection between what I'm seeing and the sound I hear, other than theoretically. Same with the keyboard player. Sure, I can strum strings or plunk keys, just not in any way that that will create pleasurable music. Singing is even more confounding. I have no idea how people make their voices sound so good. Most baffling of all: I can't figure out how several people playing together create one pleasurable—magical—aural experience for themselves and for their audience.

For many years, I never worried about this. I just enjoyed the music. But the more I thought about my bad singing and my relationship to music, the more curious I became about not just making music, but also about listening to it.

I wanted to know if we are hearing what we think we're hearing when we listen to music. I even began to wonder if music is what we think it is.

I'd started out with a simple goal: I wanted to learn how to sing. But my diagnosis as an amusic set me off on a second mission to understand what was going wrong in my brain. And that led to a third odyssey to figure out what we really hear when we listen to music. I was starting to appreciate the wisdom of Yogi Berra's "If you come to a fork in the road, take it" line — after all, whenever I hit a fork, I took both roads.

Seven months after my first visit, I returned to the BRAMS lab for more tests. Since I'd continued my singing lessons, Sean Hutchins wanted to compare the new data to what he'd learned when we'd last met. I was keen to do the tests again. Even if my ability to hear hadn't improved, at least I knew what to listen for this time. So it was possible I would do better because of what researchers call practice effects: the improvement in performance on cognitive tests as people do them again and again. This phenomenon can skew the data, but at BRAMS, they don't often test people repeatedly.

Hutchins took me into a small soundproof room and sat me down in front of the slider. Playing it definitely seemed easier this time and I didn't feel as panicked as I tried to find the note on the strip. Next up: matching my own voice. Last time, just recording *baa* at five different pitches was a challenge. I'd blurted out my *baas* blindly. But after months of singing lessons, producing a range of tones with confidence didn't seem that hard. And after he'd randomized the notes and asked me to sing them back, I had an easier time distinguishing between, say, the high and the medium notes, even if that didn't necessarily mean I could match them more accurately. This time doing the experiment was a more pleasant—less frightening—experience. Hutchins also pointed out that vocal training helps people support their voice by improving their breathing. He's tested people who could barely sustain a tone long enough. "It really does require a different type of physical co-ordination than speaking," he said. Singing was starting to feel less like a foreign language I couldn't recognize, and more like a foreign language I didn't know well enough to say much more than *bonjour.*

Weirdly, some people find it easier to sing entire songs than to match individual pitches devoid of musical context. Although both of us knew I could neither pull off a whole song nor match discrete pitches, the next experiment involved singing. "So the first song that I want to record is 'Happy Birthday.' Sing to me," he said. "Sing it the best you can, but you're not on stage here."

"You're not going to put this on the Internet, is what you're saying…"

"No, I think that would violate a few of our policies."

I was happy I'd been able to get a chuckle out of him, but what he asked me to do next really should violate the Geneva Conventions: sing the same song without lyrics, just the syllable *baa.* Still, I did it. Badly.

"We'll do it one more time," he said.

I did it again and said, "I kinda rushed through that one."

"That's fine. That one's always harder when you have to remember it without the lyrics. Are you using a strategy there?"

"No. Is there a strategy?"

Apparently not one that really works, though some people come up with different tricks they try to use. He didn't tell me what they were.

Next, he asked me to sing a song I'd worked on in my lessons — two versions with the lyrics, two versions without. I picked "A Good Year for the Roses" and, after I finished, said sorry.

"That's fine. That's the hardest test I'm going to give you."

I wish singing never felt like a test at all. The fact that some famous and successful people have been fellow sufferers is small comfort. In what is surely a cosmic poli-sci joke, given how often we complain that they are tone deaf, some prominent politicians have likely been amusic. Ulysses S. Grant, the eighteenth president of the United States, once quipped, "I only know two tunes: one of them is 'Yankee Doodle' and the other one isn't." If he wasn't kidding, his melody memory was dismal.

And although Theodore Roosevelt, the twenty-sixth U.S. president, enjoyed singing, he wasn't any good at it. "Nearer, My God, To Thee" was a fave, but James Amos, his valet, said, "I had to laugh because it would be hard to imagine anything further from 'Nearer, My God, To

Thee' than the tune Mr. Roosevelt was singing. I've heard him sing that tune scores of times, but never anything like the real tune and never the same way twice." Hillary Clinton, who would like to be the first female American president, may be tone deaf as well, if her 2007 rendition of the national anthem is any indication.

Perhaps guerrilla leader Che Guevara figured being unelected would allow him to escape the political disorder called tone deafness, but it made no difference to the musical one of the same name. Che was on the nine-month South American journey he later wrote about in *The Motorcycle Diaries* when he celebrated his twenty-fourth birthday in Peru while working in a leper colony. The booze-fuelled party raged until 3 a.m. and, according to biographer Jon Lee Anderson, whenever the band played a tango, his friend Alberto Granado would poke the amusic Che to let him know. At one point, Granado nudged him about something else and the birthday boy unwittingly rose for a mambo instead. "He took to the floor, doggedly dancing a slow and passionate tango while everyone around them jiggled to the *shora*," Anderson writes. "Alberto was laughing too hard to correct him."

Presidents and revolutionaries aren't the only ones; sometimes even scientists can't hear the way they want. Darwin — who, let's not forget, argued that the more musical the man, the more partners he attracted — attended concerts and loved music, especially when it meant listening to his wife, an accomplished pianist. But he described himself as "utterly destitute of ear" and incapable of recognizing tunes. Meanwhile, music wasn't better than sex for

Sigmund Freud. The Austrian neurologist and psychoanalyst rarely listened to music and didn't write about it. But in *The Moses of Michelangelo* he ponders the power that art, especially literature and sculpture, has over him and how he likes to spend a long time trying to figure out its sway. "Whenever I cannot do this, as for instance with music, I am almost incapable of obtaining pleasure. Some rationalistic, or perhaps analytic, turn of mind in me rebels against being moved by a thing without knowing why I am thus affected and what it is that affects me."

In *Musicophilia*, Oliver Sacks notes with surprise that there are no references to music in *Principles of Psychology* by William James—and only rarely does it show up in the fiction of his brother, Henry James. Diagnosing historical figures is probably not wise, though. After all, some of these people might have been musically anhedonic rather than amusic. Or, as Sacks suggests about the James brothers, maybe after having been deprived of music when they were young, they developed a kind of "emotional amusia."

If Henry James was tone deaf, he wasn't the only writer who suffered from it. In his memoir *Speak, Memory*, Vladimir Nabokov admits, "Music, I regret to say, affects me merely as an arbitrary succession of more or less irritating sounds... The concert piano and all wind instruments bore me in smaller doses and flay me in larger ones." Perhaps because Nabokov's prose was so melodic, Sacks wonders if he might have been joking. But rather than a laughing matter, the inability to enjoy music was probably a deep regret for the Russian-American author, especially after his only son, Dmitri, became a professional opera singer.

At least one scribbler—Charles Lamb, an English writer who was a contemporary of Romantic poets such as Coleridge and Wordsworth—had some fun with his deficit. "A Chapter on Ears," which appeared in his *Essays of Elia* collection, is a meditation on his affliction. "I have no ear," he begins. But after making it clear that he does have "those exterior twin appendages," he writes: "When therefore I say that I have no ear, you will understand me to mean—*for music.*" Not only does he admit to having worked on "God Save the King" his whole life, without ever being able to manage it, he isn't sure what a musical note is and can't tell a tenor from a soprano. "I have sat through an Italian Opera, till, for sheer pain, and inexplicable anguish, I have rushed out into the noisiest places of the crowded streets, to solace myself with sounds, which I was not obliged to follow, and get rid of the distracting torment of endless, fruitless, barren attention!"

After lunch and an espresso in Isabelle Peretz's office, I joined Marion Cousineau, a post-doctoral fellow. Growing up in France, she'd learned some piano when she was a kid, then played a little saxophone in jazz bands. The past few years she'd been teaching herself guitar and bass and doing a bit of singing. But her formal training wasn't in music; it was initially in computer science. While doing her master's in cognitive psychology, she took a course on perception. The professor asked many questions about how we perceive sounds that she'd never considered. She realized the answers were not so straightforward and became fascinated by psychoacoustics, the study of the way we perceive sound. Then she met Peretz, who was interested in using the techniques of psychoacoustics to explore amusia. Even though I didn't understand everything Cousineau was talking about, I liked listening to her. She was so clearly into it, sprinkling in words such as amazing, wonderful, and thrilling with evident enthusiasm.

Before my trip to the lab, Cousineau had asked me to redo the online test, which had changed slightly since the first time I'd done it. The rhythm block from the Montreal Battery was now one of the three parts. I listened to a snippet of music and had to indicate whether or not it had an incongruous pause. I scored a 96. (I don't mind saying, I was damn proud about that.) But a well-below-random 42 on another block, which asked if a melody had a wrong note in it, brought my average down to 69 percent. Another fail. For me, hearing an unnatural pause seemed too easy, but picking out one wrong note had me wondering, *How could anyone do this?*

In the lab, Cousineau had an afternoon's worth of tests, which to me were just more confirmation that I wasn't hearing what I was supposed to hear. Of course, she already knew that. She was studying other things, such as how amusics process the sound they hear.

By late afternoon, I finally escaped. The whole day had been a misery. I'd spent most of the last eight hours cooped up in a tiny booth, staring at a computer screen with headphones on, failing test after test, first with Hutchins and then with Cousineau.

I left the old convent, plugged my earbuds into my

exterior twin appendages, and moped my way to the Metro station. "Screw the singing lessons," I resolved. "I give up."

In between the trials, Cousineau was patient enough to answer my questions. I had a lot of them. Many were about what she was trying to discover with her tests, but some were about me. For example: How could I love music if I couldn't hear it properly? She pointed out that while control subjects in the lab can usually hear differences as small as a fifth of a semitone, no amusics can. But in Western music, the smallest interval, or distance between pitches, is the semitone. And I wasn't so badly off that I couldn't hear that. I'd probably have more trouble with, say, Indian music, which uses tiny intervals. Since I mostly listen to Western music, perhaps that was a bit of good news.

Naturally, the more I learned about my deficit, the more I found myself pondering what I listened to. Amusia hadn't affected my ability to love music, but maybe it affected what I loved.

I'm sure many people would consider my taste in music fairly adolescent. And they wouldn't be wrong—I am an unrepentant middle-aged teenager. My collection includes rock, country, R&B, blues, and a little reggae, though admittedly not much heavy metal, hip-hop, or electronic. And contemporary pop music doesn't do anything for me and hasn't for a few decades. After Apple released the iPod in the fall of 2001, I begged and wheedled Carmen in the hopes of getting one for Christmas. Somehow she consented to this extravagance, and that's how I became the first person

I knew with a device that would soon become something just about everyone owned. For a while, I'd have said that if I could take one thing out of my burning house, it would be my iPod. Several years later, when Apple's stock price went from the doldrums to an investing darling, a friend in the financial industry chirped me: "You were smart enough to get an iPod early, but were you smart enough to buy Apple stock?"

In the early days, some people liked to do iPod audits: They'd scroll through the artists on your device to get a sense of what you listened to. (This was more feasible with the original 5 gigabyte version than the sadly now-defunct 160 GB iPod Classic.) When a former student performed one of these audits of my mobile collection, she saw a playlist called Pop. She checked it out, but was surprised and disappointed that she knew none of the songs. My idea of the perfect pop song was—and is—"Tempted" by Squeeze; hers was boy bands and stuff like "...Baby One More Time" by Britney Spears, music that's completely lost on me.

I imagine that said something about our relative ages, but when it comes to more adult fare, I owned hardly any world music and, until a few years ago, embarrassingly little jazz beyond Chet Baker. Nor had I had acquired any opera or classical until the fall of 2009, when I decided to take a couple of night courses. Introduction to World and Early European Music covered Medieval to Baroque, from Gregorian chant to Handel. I didn't know the music or the related material when I started the course (and, frankly, I'm still fuzzy on some of the terminology). When I handed in the midterm, which required defining terms and identifying snippets from a listening list, I hoped I'd eked out a B but feared a C was possible. I hadn't written an exam in a long time, and it was a university course. When I teach, I expect a lot more work from my students than the two or three days of cramming I put in before the test. Somehow, though, I'd nailed it: 88 percent. Even more surprising: the rest of the class did poorly, with an average in the 50s or 60s. Clearly, some of my colleagues had failed.

"You're a hard marker," a woman moaned to the instructor.

"No, I am an accurate marker," he responded. (I really must remember that line, I thought.)

Later, as we filed out of the classroom and into the school's institutional hallway, a middle-aged woman asked me what my grade was.

"I did well."

"No. You're being too quiet," she persisted. "Tell me what you got."

"Eighty-eight."

"Why are you taking this course if you already know it?'

"I don't know it," I defended myself. "I studied."

"I hate you."

I felt like joking, "As if you're the first woman to ever say that to me." But I just walked away.

Good grades in that course and its sequel, called Introduction to Classical Music, weren't the point, though. I never heard much classical, opera, or jazz growing up. Although my wife has some classical and jazz CDs, she mostly played them when I wasn't around. Maybe, I figured, if I learned something about these genres that I'd long considered boring, I might like them. And it turned out that I did enjoy a great deal of it. I'm now a big fan of Beethoven, for example. I suppose that's a bit like saying you like pop music because you like The Beatles, but I always feel a little quiver of excitement when his Symphony No. 5 comes on my iPod, especially when it follows something loud and punky. I don't want to dance, but I do feel the urge to move my head and at least one arm as the orchestration swells and swoops.

That summer, after seeing an exhibit called We Want Miles: Miles Davis vs. Jazz at Montreal's Musée des Beaux-Arts, I added some of his songs to my iPod (courtesy of Carmen's CD collection). So I can only assume my previous antipathy toward these grown-up genres was due to a lack of exposure rather than an inability to process what I was hearing.

As dispirited as I felt after the first day, I had promised to return to the lab the next morning.

When I arrived, Cousineau had more experiments for me. Back in front of a computer in the small soundproof room, I listened to a series of sounds and combinations of sounds. There were three groups: saxophone, piano, and human voice (but not singing or talking). All I had to do was indicate if I liked what I heard or not by hitting a key to give it a score on a nine-point scale between minus four and plus four. I did as she asked. After I'd finished the first block, I told her that I liked most of the sounds, except the human voices crying. "Is that a problem?"

"No, it's not," she said. She explained that she was studying how amusics perceive consonance and dissonance. Consonant chords or intervals are, to someone who processes music normally, pleasant sounding and create a sense of calm or resolution. A dissonant combination of notes sounds unpleasant, unstable, or out of place, creating a sense of tension. In a piece of music, consonance will usually follow dissonance to offer the listener release from the tension.

Researchers know that's how the general population responds to consonance and dissonance, but Cousineau wanted to find out how amusics react. "I'll tell you more about it when we're finished," she said. And so I listened to the next block, which consisted of synthetic sounds — funny keyboard sounds, she called them. The first group featured different pitches and the second different timbres.

When I'd finished, she asked me: "How did it go? You liked some, you disliked some?"

I'd rated almost all of the sounds between neutral and positive. "I guess they just weren't sounds that I disliked. I hope that that's still useful for you." As long as I'd responded with a range of answers, the data would be useful, she explained. (But that wouldn't have been the case if I'd hit the same key for every one.)

In the last block, the computer presented me with three artificial sounds. Two of them had the same structure, making them seem as if they'd been played by the same instrument. The third had a different structure, so it sounded like a different instrument. That was the one I had to pick by pressing one, two, or three on the keyboard. This exercise wasn't about pitch, it was about timbre, or the distinct sound each instrument has. Timbre was a term I was hearing more and more; but I still had no idea how essential it is to the way we hear music.

Once I'd finished all the trials, Cousineau explained that amusics have trouble perceiving dissonance. One previous study had asked subjects to listen to short excerpts of classical, lively music—played both normally and again with the piano player's left hand shifted by a semitone, which made many of the chords dissonant. For most people, the shifted-chords versions made for unpleasant listening, but amusics tended to like both versions about the same.

Because dissonance helps create tension in music, film score composers use it to great effect. When I'm watching a film, I do hear—and feel—it. That may be because the dissonance is more pronounced, or perhaps I'm responding to other cues such as rhythm and timbre. Or even non-musical elements. "There may be other things, especially in movies, that are all communicating the same emotion," said Cousineau. "They want you to be happy or sad or scared, so they put all of the cues going in the same direction."

I was still a bit baffled by my disorder, but perhaps I could find an advantage to it: While a normal listener might hear something and immediately notice that it sounds really off, it'll still sound okay to me.

And, yet, there's plenty of music that I don't like. I think I listen to music the same way other people do, but I'm clearly hearing it at least somewhat differently. Cousineau's take is that "each person's way of enjoying music is different." And, yes, there likely are some details that I don't hear that other people hear. But those aspects—some subtle elements that I can't differentiate—don't prevent me from liking what I'm hearing.

Not surprisingly, trained musicians have much smaller thresholds for pitch deviations than most listeners. Mine, on the other hand, is much larger than normal. Still, since my threshold for individual pitch is less than a semitone, if she played a wrong note on the piano, I'd likely hear it, if the melody was familiar enough. "But," she said, "if I were to play on the violin and be a little bit out of key, I'm a little too high in pitch, you might not hear that."

She left me with a story about visiting New York with her father and brother the previous summer. They'd decided to go to a cello concert on a boat in Brooklyn. The cellist made "little bit out-of-key sounds," perhaps because of the rocking of the boat. "My brother, who's a violin player, and I were just"—here Cousineau made an *eek* sound and unpleasant face, indicating a form of aural torture—"but most people didn't react to that at all."

When Cousineau was finished with me, Felezeu took over. Her last experiment was an auditory test, just to make sure my embarrassing flaw wasn't due to a hearing problem. But my problem had nothing to do with my ears. Despite all the loud concerts I go to without ever wearing earplugs and the cranked volume on my iPod, my results on Felezeu's tests were above average, and I hadn't lost the ability to hear high frequencies. That's unusual for someone my age.

"Finally," I said, "after three days in this lab, some good news."

I was about to get a bit more. Just before I left BRAMS, Hutchins and I found a couch and sat down to discuss the results of the tests he'd given me the day before. While playing the slider had seemed easier than it had in March, my ability to match the target pitch hadn't improved. But the singing test was a bit more encouraging: he'd identified "moderate improvement" from my first visit. Actually, the numbers weren't that impressive. I'd matched five out of twenty, up from three out of twenty in March. Most people match about eighteen out of twenty, but it was a glimmer of hope. He'd also measured how off I was on the ones I missed and found "modest improvement" there, too.

I asked if he thought I should continue my singing lessons. "You do have pitch perception abilities that are below what you'd find in most people," he said. "When you go through any music lessons, you're not able to use what you consciously hear to influence your singing as much as other people might. So you'll need more practice to improve your singing than someone else." He didn't want to oversell my potential, but he thought I should go for it.

As I left the lab and walked down the hill to the Metro station, I felt much more optimistic than I had the day before. I pulled out my phone and emailed Barnes with a report on what I'd learned at the lab. He responded: "This is good news about your forward motion . . . it's scientifically tested. Makes your coach happy."

"Blue Highways"

Psyche Loui started to become excited as she looked at white matter on her computer. We sometimes talk about our brains as our grey matter, even going so far as to use it as a synonym for intelligence. But we have white matter, too. It's in our central nervous systems, and the more scientists learn about that stuff, the more they realize what a crucial role it plays in our bodies. Both the white and the grey are made of neural tissue, but you can think of white matter as the fibre optics that connects different grey matter regions.

As a cognitive neuroscientist, Loui studies how the brain enables behaviour. She was looking at white matter to try to figure out what's really going on in the tone deaf noggin. From the work of other researchers, including at BRAMS, she and her colleagues at Harvard Medical School's Music and Neuroimaging Lab knew there were grey matter differences in areas that are important for auditory perception and for sound production. She believed the arcuate fasciculus—*arcuate* means shaped like an arc or bow and *fasciculus* refers to a bundle of white matter—was the culprit. A major brain pathway, it connects the temporal lobe (home of the primary auditory cortex, which plays a role in the perception of sound) at the side of the brain with the frontal lobe (home of the sound production region) at the front of the brain.

The pictures Loui was studying had been captured in a magnetic resonance imaging machine. But they were not the high-resolution structural scans you might see if you searched the Internet for pictures of brains. These were created using a technique called diffusion tensor imaging, which tracks how water moves in biological tissue—in this case, in the white matter. Diffusion tensor imaging scans can indicate which parts of the brain are connected to each other and how strong the connections are. As we learned in grade school, our bodies are mostly water, but it's not as though water is sloshing around in our melons. It moves really, really slowly, much more slowly than blood flows. If two brain regions are well connected, then water tends to travel more efficiently between them than if the connection is poor. Diffusion tensor imaging provides a picture of how well water moves and in what directions. Then researchers can use computer algorithms to reconstruct those pathways, providing a good sense of a brain's white matter structure. As Loui explained, "It's like making a map of the city based on where the cars are going."

Armed with special software, she and her colleagues had just figured out how to identify the arcuate fasciculus in diffusion tensor imaging scans. As she sat at her computer going through the brains, she could see four branches of the pathway, two on the left and two on the right, in her control subjects. But in all the amusic subjects, she could see only three of the branches. The right superior branch wasn't visible. Then she started looking at scans the lab had collected for other studies and came across one that showed only three of the four main branches of the arcuate fasciculus. This person, she predicted, was tone deaf.

Although it was the weekend, Loui emailed the subject and asked her to take the lab's online pitch discrimination test. As soon as the woman replied with her results—which suggested she was amusic (a finding that was confirmed with more detailed testing in the lab)—Loui figured she had the evidence she was looking for. "When the email came back with the results," she remembered, "that's when I was really jumping up and down."

Later, using different software, she was able to see the right superior branch in the brain scans of amusics, but that didn't mean her hypothesis fell apart. The new technique showed that the arcuate fasciculus is smaller and not as well developed as it is in normal people. The neural pathways in the human brain create a network like a transportation network with varying classes of roads in it. Some pathways are like large highways that allow a lot of traffic to move efficiently; other pathways are smaller and let less traffic through. So a musical person may have an interstate for an arcuate fasciculus, making efficient communication between the cortices possible, while amusics have country roads.

Exactly why is something researchers are still trying to figure out, but I may be tone deaf because of disorganization. The neurons in white matter contain axons, and one possibility is that the structural arrangement of the axons within the amusic brain is more disorganized. Or it could be that the protective white matter bundles around these axons are less organized, resulting in a smaller arcuate fasciculus. Whether smaller and less organized means less active or just less efficient is still unclear. But the practical consequence is that tone deaf people can't hear themselves sing as well as others can because there is less of a connection between the part of the brain that helps us produce sound and the part that helps us listen.

After Loui explained the road metaphor to me, I began to think of my arcuate fasciculus as my "blue highway," a name for the older two-lane routes that often appeared on American road maps in blue. On a road trip, blue highways offer a more intriguing and scenic option, though a slower and less efficient one. They can be a great way to get where you're going when the going is more important than the getting. But in my brain, my blue highway was just a frustration.

After months of lessons, not to mention my visits to the BRAMS lab, I knew what it meant to be amusic. What I didn't know was why I was that way. What the hell was going wrong in my brain? Loui agreed to answer some of my questions, but she also wanted to know if I'd be interested in taking some of her tests. Would I be up for an

MRI? she wondered. Before I flew to Boston, she sent me a link to her lab's online pitch discrimination test. When an email arrived with my results, I was surprised to read, "At 500 Hz you can reliably hear pitch differences of 9.75 Hz, which means you did better than approximately 35.9% of people who took our test!"

Pleased with myself, I forwarded the email to Loui, adding, "Wow. I've done a lot of tests like this and this is the first time I've done well on one." Well being a relative term here, obviously: my results were definitely below normal. Hertz is a measurement of the frequency of vibration of sound waves, or pitch. Middle C on a piano is 256 Hz, meaning the sound waves vibrate 256 times a second. At 500 Hz, almost an octave higher, I could make distinctions of just under half a semitone. But I didn't know that at the time; all I read was that I'd done better than nearly 40 percent of the test-takers. I was delighted to score anything higher than embarrassingly bad, though I also realized I shouldn't get too excited—I knew I couldn't hear what I was supposed to hear.

Through no fault of her own, Loui tends to make the average person feel inadequate. Born in Hong Kong, she started playing piano at five and violin at seven. Her family left for Canada in 1994, as part of the great exodus before China took control of the British colony in 1997. She was thirteen when she arrived in Vancouver just in time for high school—difficult years for most teenagers, and she was no exception. So after graduation, she headed for a distant corner of the continent: to Duke University in Durham, North Carolina. She has been living in the United States ever since.

The plan for Duke was pre-med ("because that's what you do if you're Chinese-Canadian," she explained). But her love of music led her to take music and psychology courses, and she found neuroscience the most interesting part of biology so she switched to psychology. By junior year, she'd become fascinated by the intersection of the subjective (music) with the objective (neuroscience). Torn between pursuing a career as a professional violinist or going to grad school, she was intrigued when she learned that a friend had been accepted to Georgetown University in Washington, D.C., to study music and the brain. So she did her senior-year thesis on how the brain processes musical syntax by comparing the brainwaves of people paying attention to music with those of people not paying attention. "That," she said, "was the gateway drug."

Hooked on trying to understand how we perceive and produce music, Loui finished her Ph.D. in psychology at University of California Berkley in 2007. When I first met her in the winter of 2012, she was an instructor of neurology at Boston's Beth Israel Deaconess Medical Center and part of the Music and Neuroimaging Lab. Since it was Harvard, I was expecting the lab, which was on the main floor of the hospital, to be a gleaming new facility, but it was a bit dingy. The lab's director is Gottfried Schlaug, one of the leading researchers in the field, but Loui had already made a name for herself with her research on the role of the arcuate fasciculus in amusia. So it was no surprise that, in 2013, she moved to Wesleyan University in Middletown, Connecticut, to open her own research centre, the Music, Imaging, and Neural Dynamics Laboratory, or the MIND Lab.

Though Loui is clearly an overachiever, she doesn't present as one. She's enthusiastic and fun and curious about subjects beyond her own work, and the first day I met her she was dressed more like a student than a faculty member. Although she'd opted for life as an academic and researcher, Loui kept at her music, playing in an avant-garde jazz band when she was at Berkeley. In Boston, she's a violinist in the Longwood Symphony Orchestra. Made up of medical professionals in the city, it performs four or five times a year, raising money for medical or social causes. She's also in a string quartet called Folie à Quatre, a riff on folie à deux, a term for when two people infect each other with their psychosis. "We're four people infecting each other with our psychosis," she said. "And the fun part of it is, in addition to me, there are two psychologists and one clinical psychologist so we're all interested in the mind." They mostly play pieces by composers who had some psychological or neurological abnormality or disorder (German composer Robert Schumann, for example) and often combine their performances with lectures on mental illness and creativity.

Despite being an accomplished musician, Loui's tastes go beyond classical to genres such as folk, classic rock, and Asian pop (for reasons of nostalgia, she explained, rather than the sophistication of the musical structure). And she doesn't mind listening to bad singers for her job. "As a classically trained musician, everything has to be just so," she said. "But listening to amusics singing 'Happy Birthday' has made me a lot more open minded about what sounds you can accept as being music. As long as people are enjoying themselves, I think it should be encouraged."

Loui's research doesn't just focus on amusia. She explores a range of topics related to how we perceive and produce music. That includes perfect pitch. People with this rare gift, which is also known as absolute pitch, can identify or sing a pitch without a reference note. Just one person in ten thousand has it. More common is relative pitch, which is the ability to identify or sing a note by comparing it to a reference note. Surprisingly, the absolutists are not necessarily better at hearing small differences in pitch; they're just astonishingly good at categorizing pitches. Not surprisingly, some great musicians—including Mozart, Beethoven, and Bach—were blessed with perfect pitch. Another of the gifted is Diana Deutsch, a cognitive psychologist at the University of California San Diego. According to her, "People with absolute pitch name musical notes as rapidly and effortlessly as most people name colours."

Loui believes we all have some innate knowledge of music. While she's not sure we're born musical, we're likely born with some predispositions: Day-old infants can detect a beat in music, and three-month-old infants can detect pitch changes. Still, we're not born knowing all the complexities that grown-up brains show. For example, we don't come out of the womb with a grasp of subtleties such as implied harmony, the idea that we can hear chords even if the root note isn't there because our minds infer it. So it's not like the rules of music are written in our brains. In the same way, we aren't born with the rules of language (unless you buy Noam Chomsky's Universal Grammar theory). If we were, then languages around the world would be much more similar than they are—and, I might add, more people would have better grammar. But we are born with the capacity to learn.

I asked Loui how people feel when she diagnoses them as amusic. Some of them, she admitted, were offended. "They think you're telling them they're incompetent, which is totally not what we think," she said. But once she explains to them that it has nothing to do with being deaf or having any other developmental issues, most people are quite happy that she's found an explanation for what's bothered them for most of their life. It can be a liberating experience. She's received emails from people thanking her because they'd been wondering since that time they were in the school choir and the teacher told them not to sing. "Now we've actually found a reason for that," she said. "It's real. It's not like they're lazy or not paying attention or incapable — there's actually a scientific cause for it."

Tone deafness fascinates her because a lack of musical ability in otherwise high-functioning adults is a surprising mismatch. She also thinks researchers can learn a lot from studying amusia. It's a window into the human brain that sheds light on how hearing works, how auditory perception works, and how we are able to use our voices to produce sounds. The way music works in the brain is complex. First, you have the perception of pitch, rhythm, harmony, melody, and timbre, and all the things that come out of the sound. Riding on top of the sound signals are cognitive elements such as working memory, attention, learning, and expectations. Then there's your emotional response to music and motor aspects such as bopping your head to the beat. "It's interesting to test what happens if you don't have one of those components," she said. "There's something different about brains of tone deaf people that's informative of what the concept of music actually is." She also thinks amusia research provides a glimpse into how the brain works more generally.

"So if you study the people who are abnormal," I asked, "you can learn more about the normal?"

"Yeah, it's like studying autistic children to find out why people are social."

We left Loui's tiny office and she showed me around the lab, introducing me to some of her colleagues, before taking me down to the basement for testing. Some of the tests were similar to the ones I'd done at BRAMS, but some weren't. Inevitably, some involved singing. I listened to two tones and had to decide if the second was high or lower; then I listened to two tones and hummed both of them back into a microphone. So the first was about perception and the second was about production. It turns out that some amusics can produce pitch better than they can hear it while others are the opposite. "It's quite common to be better at production than perception, and it's also quite common to be very bad at production but not so bad at perception," said Loui. "So I think it's really the mismatch that's most characteristic."

These were threshold tests, a way of narrowing down the results. After I was correct three times, the tones became closer in pitch to make it more difficult, but every time I was wrong, the pitch difference grew, making it

easier. Using this method, she could find the smallest difference in pitch that I could reliably hear or sing. Different labs
use different measures for an amusia diagnosis; to Loui, a
threshold of between half a semitone and a semitone means
you're mildly amusic, while anything more than a semitone puts you at strongly amusic. After I hummed all the
notes, she gave me my results: "Okay, so your production
threshold is worse than your perception threshold. Your
perception threshold just now was 16 hertz, which is half a
semitone, which is definitely borderline, very clearly within
the borderline," she informed me. My production threshold,
though, was larger than a semitone. In other words, I don't
hear pitch well, but I really suck at singing.

For the next test, I was to hum the tones back and then
say whether the second one was higher or lower. This one
was the most interesting to Loui. "We can talk more about
how that felt at the end of it," she said, "because I'm very
curious. I think this is the one that gets most into what's
in your head."

When I'd finished, she asked, "Which part was more
difficult?"

Hard to say. I found both difficult because I had to
concentrate on humming the tones and think of the right
answer to the question at the same time.

"In all those pairs of tones," she asked, "did you feel like
the first tone sounded the same the whole time?"

"No. A lot of the time, but no. There were some that
were a little different, weren't they?"

Nope. Wrong again. The first one was always the same,
the second was either higher or lower. "The interesting

thing is that you didn't always start singing on the right note but you always sang in the right direction." I wasn't always able to name the direction of the second note, though. Loui saw this as mismatch between the conscious and the unconscious, which fit with the prevailing theories. If the hearing parts of your brain aren't well connected to the production parts, then you can't hear yourself sing or sing what you think you're hearing. "That's the crux of the problem," she said. "I think it's pretty interesting."

I sang "Happy Birthday" for her. "Thank you," she said cheerfully. "You got all the directions right. That's representative of tone deaf people. When they sing 'Happy Birthday,' they will not get the intervals right, but they'll get the direction—so, the contour—almost always right."

"So contour's a lot easier than pitch?"

"I guess it's a prerequisite of pitch, and it's also less under voluntary control, more an involuntary unconscious process."

I'd had a couple of MRIS on my knee, which meant my head had stayed out of the machine. That made me contemplate how weird it would be if my doctor wanted an MRI on my shoulder or head, and I had to be all the way in there with all that noise. I'm not claustrophobic, but surely that bizarre mix of buzzes and bangs, clicks and clangs, would be even more unsettling—or at least louder—if you had to be all the way inside the machine. This thought occurred to me when Loui asked if I'd be willing to submit to an MRI. But mostly I couldn't help but delight in a memory of an episode of *The Simpsons* called "HOMR," in which an X-ray reveals a crayon lodged in Homer's brain (presumably dating from boyhood when he liked to shove crayons up his nose).

But this is what I learned about MRI machines while I was in Boston: you get used to them. Early on my second morning there, I met Loui at the hospital and signed a bunch of forms, including one that made it clear that the procedure might reveal some bad news about my brain. (No mention of crayons, though.) Then I changed into two of those ridiculous hospital gowns: one opening at the front, one opening at the back, just so there would be no wardrobe malfunctions. Once inside the room with the MRI, I saw where I would spend the next forty-five minutes of my life.

An MRI, or magnetic resonance imaging, scanner looks like something that should be part of a NASA mission, maybe the portal that takes the astronauts from the capsule to outer space. It features a large round cylinder with a big round hole in the middle. Just outside the doughnut hole is a long tray known as the patient table. I climbed on the tray and lay on my back as Loui and the operator gave me foam earplugs, put a helmet with a microphone attached on me, and placed a panic button in my hand so I could alert them if I started to freak out. Once I was ready, the patient table glided into the machine's hole; I felt like a giant pizza sliding into a mechanical wood-burning oven. But instead of heat, there would just be a hell of a racket.

Our understanding of how the human brain works has deepened dramatically in the last two decades. One reason:

the functional MRI, or fMRI, allows researchers to study the brain in ways other methods can't. An MRI shows us the structure of the brain, which is useful, but an fMRI reveals activity in the brain using the same scanner, and that's even more useful. Functional MRIs help us understand how our brains respond to outside stimuli by letting researchers see increased blood flow to areas during certain emotions, thoughts, or activities. EEGs (electroencephalograms), which measure electrical activity, and PET (positron emission tomography) scans, which require the injection of a radioactive compound, do capture some brain activity. But they can't do what fMRIS do: Give us a sense of what goes on in our heads when, say, we recognize a face or someone has a broken heart or jazz musicians trade musical phrases.

The MRI scanner—which uses radio signals and a really strong magnetic field to produce images of tissue and organs without using radiation—has been around since 1980, five decades after scientists figured out how neurons respond to a magnetic field. You may have had a structural MRI on an injured body part and wondered how these machines work. Well, like this: The magnetic field causes protons in your body to all line up in one direction. Then short bursts of radio waves make them spin. When the radio waves stop, the protons realign, releasing the energy they'd absorbed. Protons in different tissues realign at different speeds, and computer algorithms allow researchers to turn the data into 3-D images. That's useful when a doctor wants to know if, say, your knee pain is because you have a small tear in your anterior cruciate ligament.

In the early '90s, researchers proved that the same

technology works for studying brain activity. They gave fMRIS to people watching flashing lights or tensing their hands, then crunched the data to create images of the visual and motor cortices of the brain. While all parts of our brains are always active — I know, I was surprised to learn this, too — an fMRI helps researchers see small differences. When neurons fire, they use more oxygen, so that part of your brain needs blood to deliver more oxygen. And it just so happens that blood has different magnetic properties depending on how much oxygen it has in it. So instead of mapping tissue, fMRIS map blood flow.

"When people say, this is your brain on music, that's most likely referring to the functional activation that's due to music," explained Loui. When we listen to music, we use many parts of our brain. "The blood vessels that are subserving those neurons are really acting like a gossipy neighbour who says, Oh, look, the brain area that I'm supplying needs more oxygen now. That's exactly what the fMRI is measuring."

This discovery led to an explosion of research. Initially, fMRIS showed us which parts of our brains did what — where language perception takes place, for example, or where we store our memories. A lot of these studies confirmed what researchers already figured was the case. Before fMRIS, scientists could deduce what brain region was responsible for what by studying how damage to that region led to a deficit in a certain brain activity. Functional MRIS were much better at that. They didn't just show researchers when their deductions were right, they also showed them that the human brain is often not as simple as they thought.

As scientific understanding of the role of each brain region increased, scientists became more ambitious, and the technology encouraged them to tackle more complex questions. How do we make moral decisions? Why is it so hard to stop smoking? What happens in the minds of people who say they're not racist when they're presented with photos of people from different races?

Each year now, researchers publish thousands of brain studies based on information gleaned from fMRIS. Increasingly, these studies look at the neural circuitry. Instead of merely studying anatomical structures, scientists want to understand the connections between neurons — including those in different parts of the brain — and what role they play in different tasks. Cognitive neuroscientists are particularly interested in how the brain enables various abilities, whether it's seeing or doing math or paying attention to other people. "All these things that make our lives interesting and rich," as Loui put it. "I guess I consider myself a cognitive neuroscientist of music, so what I'm generally interested in is how the brain allows us to make music and when does the brain not allow us to make music, such as in tone deaf people."

Researchers are understandably excited about the better understanding of brain physiology that fMRIS offer them. But overreaching is a risk. If the technology could reveal brain activity in someone in a persistent vegetative state — and potentially even allow that locked-in patient to communicate by thinking thoughts an fMRI could pick up — we'd all applaud. But it would get ethically trickier if the patient took advantage of the technology to express a wish to die in a jurisdiction with right-to-die legislation. And with some research, we need to maintain our skepticism. Marketers have used fMRI data in an attempt to improve the way they reach consumers, and at least one U.S. company offers fMRI-based lie detection services to bosses who want to screen potential employees before hiring them. Meanwhile, in India, an fMRI scan "proved" that a suspect had knowledge of a murder that only the killer could have. And some studies have suggested that fMRIS could predict which prisoners are likely to reoffend. You don't need to be a fan of dystopian science fiction to imagine how wrong some of these ideas could go.

Any excitement over the technology needs to be tempered by an understanding of its limitations. An fMRI doesn't directly measure the activity of neurons; it just shows the blood flow in the brain. That blood flow keeps changing and will be affected by age, medications, and other factors. Even when we are at rest, our neurons are active, so just because blood flow to a brain region doesn't increase when we do something in an MRI machine doesn't mean that part of the brain isn't active. One possibility is that the neuronal networks that are active at rest are different from the networks that are active during a task, so the total amount of oxygen the region uses remains the same.

Dealing with the data presents more opportunities to get it wrong, especially since one scan can generate millions of pieces of information. Without proper statistical analysis, researchers can report "noise" as good data. Neuroscientist Craig Bennett and his colleagues at the University of California Santa Barbara wanted to make this point, so they used an MRI to show "brain activity" in a dead salmon. Other researchers complained that this was an unfair attack because any scientific research will produce bad results without proper statistical analysis.

Finally, neuroscientists such as Robert Shulman, professor emeritus of molecular biophysics and biochemistry at the Yale School of Medicine, argue that too many researchers are asking questions that are so complicated (about morality, personal values, and so on) that they don't have simple answers. "While there are marvelous results to be gained from careful research on brain metabolism and blood flow as measured by fMRI," he writes, "there is little to be gained by making assumptions about the human mind."

The failure to resist the temptations of the technology has led to some dubious studies. One claimed Republicans are driven by fear more than other people because when they played a gambling game, they had more activity in the amygdala. It didn't matter that the amygdala is also connected to desire and reward, so the brain activity may have had nothing to do with fear. But pro-fMRI neuroscientists counter that the technology, the data analysis, and the design of studies have all improved. And it would be a mistake to dismiss the procedure as a space-age version of phrenology, a pseudoscience popular in the early 1800s that claimed to reveal people's character based on the shape of their skulls. After all, fMRI technology has been responsible for many worthwhile developments, including helping to change the social perception of mental illness. And it offers many intriguing possibilities for the future, including the potential to identify illnesses such as Alzheimer's before patients experience any symptoms. I'll admit I knew next to nothing about fMRIs when the patient table I was on slid into the machine in Boston. And I certainly didn't know about any controversies surrounding the way some researchers used the technology. But Loui didn't seem to be up to anything nefarious, and I was game for playing a human pizza, especially if it might reveal something about my broken brain.

Loui and the operator went to the adjacent control room so we could begin. When Loui did the initial study looking at the arcuate fasciculus, subjects didn't have to do anything when they were in the MRI. I did. This experiment required that I listen to sounds as well as sing. Strapped in, on my back, and surrounded by a massive hunk of metal, doing anything, especially singing, was awkward. And I may have sung worse than I ever have. Or maybe I just sang my usual amount of bad.

Loui had taken me through a few dry runs in the lab the day before. The basic task was to listen to a tone and then either match the pitch with a hum or, for the control condition, just breathe. The cue to sing sounded like a little burst of wind; the cue to stay silent wasn't that different: a high burst of wind, almost a *shh*.

"Why are the two sounds for hum it and don't hum it so similar?" I asked.

"To make sure that subjects are paying attention."

"That's smart. Devious, but smart."

While I was to keep my eyes closed so the results weren't skewed by any visual activity, she didn't want me to fall asleep. She admitted that when she'd participated in her

own experiment, she kept feeling sleepy. "The most important thing is that you're not claustrophobic," she said. "We
turn off the scanner background noises while we're running the scan so you're listening to sounds in as much quiet
as possible, but it's still loud."

She was right about that. The study I was part of was
looking at vocal control. By playing my own humming
back to me while I hummed—sometimes exactly as I'd
produced it, sometimes altered to be higher or lower in
pitch to trick me into thinking that I was singing out of
tune—she could measure how I compensated. Here's the
surprise: when tricked into thinking they're singing out of
tune, most people will compensate by singing in the opposite direction. If they're to hum back an A, but are being
tricked into thinking they're singing a B-flat, most people
will compensate with a note closer to a G-sharp. They do
this so they can hear themselves sing in tune, even if they're
not singing in tune. While I was in the pizza oven, she also
did a diffusion tensor imaging scan so she could construct
a model of my brain's pathways. That part was easy for me:
all I had to do was lie there and not move my head.

This study, Loui hoped, would show that amusics don't
use the perception and action regions of their brains as
much as other people, so they don't correct themselves
when they sing out of tune. In other words, they can't make
use of the auditory feedback they get from their own voice.
The crux of the problem might have to do with awareness
of pitch more than just passive perception of pitch. Not
surprisingly, teasing apart passive perception of sound from
awareness of sound isn't easy. But if Loui is right, it might

not be as simple as tone deaf people can't hear small differences in pitch. They might implicitly, or unconsciously, hear the difference but not have conscious access to or awareness of that information.

Or some of them, anyway. The emerging consensus, according to Loui, is that tone deafness isn't a single deficit so much as a grab bag of different symptoms. So there are subcategories of amusics: for some people, for example, the problem may be mostly a memory deficit; for others, primarily a learning deficit; and for others, largely an awareness deficit.

A few days after I returned home, Loui emailed me three scans of my brain. Seeing them made me feel weird. Sure, the pictures looked kinda cool, but I had no way of knowing what they meant or showed. I thought about posting them to social media, but then worried that people would see them and wonder, Oh, my God, what is wrong with that guy? Or think, That explains so much. Not that I could think of anyone I knew who would be able to tell anything from a picture of a brain, but your mind loses confidence when you know your brain is broken. At least Loui hadn't found any crayons.

Even better, she isn't without hope for my ilk. Loui believes music training may help enlarge our neural pathways, though I was a bit alarmed when she talked about stroke patients spending up to five hours a day retraining their brains. But the best bit of news I heard while I was in Boston was when Loui told me that she wasn't surprised to hear I love music — in fact, one of the best parties she's ever been to was DJed by someone she'd tested as amusic.

How We Hear Music

"Cum On Feel the Noize"

As a university student, Frank Russo snagged a job looking after special guests at Toronto's Ontario Place. One day, B.B. King came to play the Forum, the park's music venue at the time. A fan and a musician himself, Russo took the opportunity to chat with King before the show and to watch him between sets. King's reserved off-stage demeanour surprised the teenager because it was startlingly at odds with the legendary blues guitarist's on-stage style, which featured an arched back, shaking body, and frenzied facial expressions. Fascinated by how this calm, cool character transformed into a visually expressive performer, the nineteen-year-old started to wonder about the visual side of music: Were those theatrics really necessary? Was the emotion real or was it just a performance? The experience made him curious about what exactly is going on with our intense emotional responses to music.

Today, Russo is a Ryerson University psychology

professor and the director of the school's Science of Music, Auditory Research and Technology Lab (or SMART Lab, because these places all seem to need a cute acronym). Not that B. B. King changed his life or anything—there were lots of steps to his current job. The 1980s were "a lot of time spent listening to The Smiths, and my family thought I'd kill myself." But he also sang and played guitar in bands, moving from acoustic music to synth pop not dissimilar to hair bands such as Duran Duran and Depeche Mode to heavy industrial beats over the course of three or four years.

"Music, for me, was a lens to understand myself, other people, how we think, how we remember things," he said. "Eventually, it came to my attention that there are people studying the psychology of music, and it was pretty easy to get into that once I figured it out." But he likes to tell the B. B. King story, and it's a good, easy-to-grasp illustration of the way he thinks about music and why he finds it so intriguing.

Perhaps, I figured, he could help me answer the question that had gnawed at me ever since Isabelle Peretz had been so surprised to learn of my love of music: If I'm not hearing what I'm supposed to be hearing, then what the hell am I hearing?

When I first contacted Russo, I knew a bit about his work, but I didn't know how much we had in common. We both love music, playing hockey, and going on canoe trips. Nor did I know that he was about to start a study examining people who have normal pitch and rhythm perception, but are indifferent to music. Just the opposite of me, in other words. *This is bizarre,* he thought when he received my email. *I need to talk to this guy.* "It did not really occur to me that folks like you might exist," he emailed me back. But my case did mesh with the working hypothesis of the study he wanted to do.

Russo believes that not everyone has the same ability or willingness to be absorbed by music, and that music absorption—or the tendency to become lost in it—and musical aptitude are two separate things. Like many of his colleagues in the field, he also has varied research interests, including physiological responses to music such as goosebumps, sweating, and increased heart rate; music therapy for Parkinson's disease patients; music training for children with cochlear implants; and the non-auditory aspects of music.

When I sat down with Russo in his lab early in 2012, it was the first time I'd talked to him, but it wasn't the first time I'd heard him speak. Three years earlier, I'd gone to his lunchtime lecture called Discover Your Hidden Diva: Electromyographic Evidence for Imitation in Perception of Song, a title that made slightly more sense to me when I left than when I arrived. He started his talk by playing a piece of music and asking the audience what we thought it was. To me, it seemed mournful but it wasn't a dirge; in fact, there seemed to be something upbeat about it, too. I wondered if it might be some kind of funeral march. But the small classroom was full of faculty and students from the university's music and psychology departments, people who knew what they were talking about, so I wasn't about to say so out loud. Turns out I was right, though—it was a funeral march from northern Thailand. Russo's point was that emotion from music crosses cultures because it involves movement, and patterns of emotional movement are universal. "Music," he said, "is more than sound."

Now I was sitting in his lab explaining myself to him. I was barely into my tale of tone deafness before he interrupted me: "Sorry," he blurted out, "I'm restraining myself from wanting to sign you up for all kinds of research and figuring out what we can do with you."

As a journalist, I've spent most of my life asking other people about themselves and their work. Now all these scientists wanted to study me as much as I wanted to study them. I always worried that one of the researchers would start pulling out probes or launch into *A Clockwork Orange*–style experiments on me. Fortunately, Russo wasn't threatening to do anything of the sort. Yet. "We can talk about that later," he said. "I can percolate and come back to you."

What we hear when we listen to music is more com-plicated than most of us imagine. In the West, we tend to think of music as mostly about pitch and rhythm, especially pitch. But there's a lot more going on. Even the pitch-impaired can recognize songs within a fraction of a second based on the distinct sounds of the voices or instruments. So while I may be weak on pitch and meter, perhaps I make up for it by being strong on rhythm, timbre and contour (or, as Russo puts it, the movement implied by the music). "I'm sure you're hearing things in the music," Russo reassured me at our first meeting. "They're just not the things that we tend to think about in the Western classical tradition, in the science of music."

Maybe how I hear music isn't even all about hearing. A big part of Russo's work is understanding how our different senses are involved in the way we perceive music. He believes music, emotion, and movement are all connected—the word *emotion* comes from the Latin *emovere,* for move out, remove, or agitate—and music implies motion through changes such as those in pitch or tempo. "Paintings and photographs may be compelling for an instant, but few static arts sustain interest over time," writes Gary Marcus in *Guitar Zero.* "People are willing to pay so much more for views of lakes or oceans in part because they are always changing; any static view eventually loses its hold. Moving scenery keeps us engaged, and music is always moving."

We associate certain emotions with certain movements: slow and heavy suggests sadness, while intense and abrupt suggests anger. Russo admits that although his ideas about movement are as old as the Greeks, they are not the consensus view among his music cognition colleagues. "My big hypothesis about the emotional response to music—and emotional response is not the only reason people listen to music, but I think it's a really big one—is that what we're perceiving when we perceive music is movement on some abstract level," he said. "And you can see the movement, you can hear the movement, you can also feel it."

So when we think experiencing music is all about *listening,* we are wrong. First of all, there's also a lot of visual information. We can see rhythm, for example. With drummers that's obvious, but when we watch bands perform, we can also see the guitarists tap their feet or sway as they play. They probably do this to help themselves keep the beat, but it also helps listeners understand the framework of the music. And we can see pitch changes in the singer. When we watch someone sing, we tend to focus on the mouth, but the movement of the head and eyebrows also provides emotional information. Higher pitches, for example, may be accompanied by a wider mouth and raised eyebrows. Even the movement of orchestra conductors, who don't produce any sound themselves but shape it by controlling the phrasing and dynamics, can also influence what the audience hears. Not only do conductors often provide a robust visualization of the beat, according to Russo, but they also convey emotional aspects of the music through their gestures and facial expressions. In *How Music Works*, David Byrne notes that his Talking Heads bandmate Jerry Harrison has produced a lot of first albums for other groups. When he started doing this, he was surprised that they often had problems playing in time once they were in the studio. Not that they were sloppy amateurs, Byrne says, but in small clubs and even bigger halls, the visual aspect of performance masked the "lurching and shaking." When I read this, I had to email Russo to get his take. "This is really interesting," he responded. "Visual synchrony contributing to perception of auditory synchrony. I'm sure that he is on to something."

We also feel music. If I pluck a string on a guitar, I will make the string vibrate and that vibration will travel to your ear and you will hear it. If the vibration is strong enough, it will travel through floorboards; if you have your hand on a table, you can feel the vibrations from that guitar string. So you can feel pitch changes through vibration. And anyone who's been to a dance club knows what it's like to feel thumping bass. In fact, deaf people can dance to music by sensing the bass. Byrne points out that we feel low frequencies, especially in our torsos, as much as hear them. "Beyond any audible and neurological apprehension of music, in the disco environment it was pummeling and massaging us physically," writes the man responsible for so many songs I love to dance to. "These frequencies are sensuous, sexy, and also a little dirty and dangerous."

According to one legend, Beethoven, having started to lose his hearing in his late twenties, cut the legs off his piano so he could better feel the vibrations through the floor. That may or may not be apocryphal, but percussionist and composer Evelyn Glennie performs barefoot to hear better. Aside from a successful solo career, she has played with both symphony orchestras and contemporary artists such as Björk. She also happens to be profoundly deaf, and has been since she was twelve. After a long time trying to avoid discussing her deafness with journalists, she discovered that that didn't mean they wouldn't talk or write about her, it just meant they'd get a lot of the facts wrong. So in 1993, she wrote "Hearing Essay," which she hoped would set the record straight.

"Hearing is basically a specialized form of touch," the essay explains. "Sound is simply vibrating air which the ear picks up and converts to electrical signals, which are then interpreted by the brain... For some reason we tend to make a distinction between hearing a sound and feeling a vibration, in reality they are the same thing." Glennie also tells the story of how as a young student, she placed her hands against the wall while her teacher played the timpani. "Eventually I managed to distinguish the rough pitch of notes by associating where on my body I felt the sound with the sense of perfect pitch I had before losing my hearing." She felt the low sounds primarily in her legs and feet and the high ones on her face and in her neck and chest.

Even deaf people who don't have Glennie's musical talent can hear music with the Emoti-chair, which Russo developed with Deborah Fels, a business school professor who runs Ryerson University's Centre for Learning Technologies. The chair allows deaf people to feel music—including high frequencies—through vibrotactile stimulation. When they sit in it, some deaf and hard-of-hearing people will, without prompting, move their hands and bodies.

The chair looks normal enough: it's black and cushioned and I sank down low when I sat in it. The stylized TAD on the headrest stands for Tactile Audio Displays, the company started by Maria Karam, who had been part of the development team as a post-doctoral fellow. Her goal now is to make the chairs a commercial product, selling them for movie theatres, concert halls, and homes.

Hidden in the cushions are loudspeaker components called voice coils that generate the vibrations. The lowest ones are in the seat and the high ones, up to 920 Hz, are near the shoulders. Although we can hear frequencies ranging from 20 to 20,000 Hz, we can feel only those between

1 and 1,000 Hz. But 85 percent of the pitches in music fall in that latter range. By delivering different frequencies to different parts of the body, the idea is to mimic the cochlea (which processes different ranges in different parts of the organ) and allow the person in the chair to feel various frequencies at the same time.

On the day I tried it, not all the voice coils were working properly because SMART Lab researchers were still doing some fine-tuning in preparation for an experiment that was to start a week later. But I sat in the chair and put on industrial headphones like the ones worn by people who work outside at airports. At first, I could hear strains of the song that was playing through the headphones, because sound resonates through the chair and the body. It was Queen's "Crazy Little Thing Called Love." So James McGrath, the lab manager, played white noise. The chair's vibrations felt like a gentle massage.

Next, I donned some earbuds attached to Russo's phone, on which he'd launched a white noise app, and then put on the industrial headphones. I couldn't hear Russo talking to me even though he was right beside me, but I could feel the song that was playing through the chair. And unlike the white noise, this felt musical. There was a discernible structure to it. I could follow the rhythm, but it was nothing like the *thump thump thump* of a concert or a dance club. I couldn't feel the emotion, but it was a cool experience, much different from the white noise massage. The song, it turned out, was "Hey Jude" by The Beatles.

Toronto's Bob Rumball Centre for the Deaf has had an Emoti-Chair for several years. Users have reported that they start to get jazz and even understand the difference between artists. The more they take advantage of the chair, the more they develop a refined sense of what music feels like through vibrations. Deaf people can determine the emotion of a piece of music, mostly driven by tempo, and though the emotion they feel is consistent with what hearing listeners feel, it's not identical. Still, they can tell if a song is happy or sad.

When Russo was using the chair a lot, he found that his sense of vibration was heightened and he started to feel things—while riding the streetcar, for example—that he hadn't been aware of. But the project also taught him about how we hear music. "It reinforced an idea that I'd had for a while that music is more than an auditory experience," said Russo, who had previously been working on how what we see influences what we hear. "There's a visual component and there's a tactile component that's part of the everyday experience of music. But like most things that we study, we pick one thing and we focus on it at the expense of everything else."

Russo long ago decided that B.B. King's theatrics really were necessary. And according to Randy Bachman, King thought so, too. "I first played a show with B.B. in the mid-'60s," the former guitarist with The Guess Who and Bachman-Turner Overdrive said after King's death, at age eighty-nine, in 2015. "He told me I 'played pretty good for a white boy,' but needed to 'make a few faces' to show that I 'felt the notes and strings I was bending.' To this day, there are many pictures of me playing my B. B. King string-bending notes and my face looks like someone is stepping on my foot."

For those of us who were lucky enough to see King play live—as I did once, coincidently at the Molson Canadian Amphitheatre, the venue that replaced the Forum at Ontario Place—what effect did these faces have on us? If we hear, see, and feel music, what do our brains do with all that sensory information? Russo believes that when our brains engage in a simulation of the movement, it helps generate an emotional response to the music. When we watch musicians perform, for example, we'll also start to sway our bodies, tap our feet, or move our heads to the beat. Those are the big, obvious manifestations of the simulation, even if we start doing them unconsciously. But we also make more subtle movements that we're not aware of. When someone sings a large interval—from a low pitch to a high one, for example—her face changes and her head moves and we synchronize with her. So if it's a happy song, we also engage the facial muscles that we engage when we're feeling happy. And that allows us to internalize the emotion the singer is trying to convey.

Russo's lab has studied what happens to our facial muscles when we watch someone sing. The muscles activate in a way that's consistent with the singer's facial movements. It's subtle, usually too subtle to observe by eye, but researchers can detect it by recording changes in the electrical potential in the facial muscles. Within about 200 milliseconds of seeing and hearing a singer smile, we will smile in response. If this automatic process goes on long enough, the smiling will begin to influence our mood. But it's not just about the face. To create sound, a musician's body, or at least part of it, has to move. And it's in the pattern of movement that we have a sense of the emotion she's trying to convey. This is especially pronounced at the opera. When we see and hear a singer make a huge pitch leap, we can imagine the vocal effort that's involved and how that might be really emotional.

To help you understand Russo's ideas of movement and music, I will (somewhat reluctantly) wade into the controversial subject of mirror neurons. Italian neurophysiologists inserted electrodes into a macaque's motor neurons to study the area in monkey brains that's responsible for reaching and grasping. According to one version of the story—there are several and the lab notes were too vague to settle anything—when one of the researchers reached for his own food, cells in the macaque's brain fired in the same way that they would have if the monkey had reached for its own food. And yet, the macaque wasn't moving.

The paper on this discovery came out in 1992, and since then many people have inferred that the same neurons exist in the human brain. Many others have tried to debunk that idea. The defenders argue that when you, say, strum a guitar, neurons fire in your premotor cortex, which is involved in the planning and execution of the body's movements. But—and here's the mirror part—when you watch someone strum a guitar, about a fifth of those same brain cells will fire. So, if mirror neurons do exist in humans, when we watch a kid ride a bike for the first time, we don't just remember what that was like, we feel that wobbly mix of terror and thrill. The suggestion is that this brain network is connected to empathy.

This theory has generated a lot of hype and wild claims. For a time, some researchers believed autism was the result of a dysfunctional mirror neuron system, though that doesn't appear to be the case. We've also seen an inevitable backlash from people pointing out that this brain system has yet to be positively identified in humans. Believers consider the discovery one of the most exciting developments in neuroscience in the past few decades; non-believers aren't just skeptical, they're downright angry (if some of the Internet rants I've read are any indication) that anybody would believe mirror neurons exist in humans.

For his part, Russo treads carefully in this debate. He notes that just because there's no solid evidence that the neurons exist in humans, that doesn't mean we don't have some kind of mirror system. Whether or not it involves mirror neurons, he believes an internal simulation system—some researchers prefer the clunkier but less incendiary term "action-observation network"—does exist in the human brain. He can say that based on evidence from his lab and from other researchers. The system helps us rapidly understand the actions of others without a lot of deliberation. "I don't know how else you explain newborn infants smiling within the first day," he said. "It's not all gas."

This system, whatever it is, plays a role in how we experience music. In one study, the SMART Lab asked some subjects to make a structural judgement (how far apart the pitches were) about excerpts of songs and asked others to make an emotional judgement (did the music convey positive or negative feelings). The experiment revealed that the mirror system was highly engaged when subjects made emotional judgements, but not engaged when they made structural ones. "A lot of listening to music is emotional," Russo said. "That's not what they teach you in music school and, generally speaking, it's not what people learn when they learn music, but I think emotion is a huge part of music." He wasn't talking only about aesthetics. "It's not just 'that makes me feel good' or 'that was beautiful,' there's also a kind of emotional communication. And some of the hardware that allows us to do this emotional communication is this mirror system."

Singers produce melody, for example, by manipulating their vocal apparatus. You can hear and see that movement, but you also feel it because of the internal simulation going on in your brain. "So if I'm feeling down in the dumps because something's not working for me professionally, my relationship's on the rocks, whatever it is, I'm moving slow, I'm feeling heavy," explained Russo. A piece of music can convey that movement in both the pitch contour and in the way the rhythm and tempo change and unfold. "All that movement is internalized and represented in the brain and, on some level, we are moving in a way that music was created."

That includes the fine movements that are necessary to create pitch changes, but also some gross movements such as musicians swaying to the beat. They may not do it on every beat—they may move every fourth beat, at the level of the bar—but that helps us to understand the emotion in the music. So the way the brain maps sensation allows us to feel the emotion that led the performer to move that way in the first place.

Some researchers believe that when it comes to music, the mirror system is both visual and auditory (and the simulation is strongest when you can both hear and see the movements). So, for example, when a piece of music moves in a lively way with big pitch jumps and accelerations in tempo, it conveys a sense of movement that you've experienced when you're feeling manic. And when you hear someone sing, your brain silently sings along.

Russo doesn't think the mirror network is the only way we perceive emotion in music. Lots of other things are going on, including the lyrics and personal associations such as the they're-playing-our-song phenomenon, that colour our emotional response. But he's convinced that movement is an essential part of the experience.

This thinking takes us a long way from the traditional focus on precise pitches. "We have the staff, we have music notation, we have intervals—these are the rudiments of music that we tend to think about as important and they are, if you want to teach someone how to produce music in this tradition," Russo said. "But I don't know that they're important to our emotional response to music."

People who can hear pitch well would hear some movement in it, but even people like me can get a sense of the movement through the rhythm and the gross contour, or shape, of a piece of music. "The magic in a pop tune is not going from C to F to G to C; it's about the contour, the gross pitch contour that brings us from that first C back to that final C," Russo said. I may also be picking up on timing nuances. He cited Amy Winehouse and some jazz musicians as being famously late, or off the beat. (Being early suggests anxiety or tension, while being late suggests relaxation.) "I suspect that you can hear all that and it does something for you. It turns emotional dials for you."

Music, Russo said, is "massively overdetermined," which is to say it's full of redundant cues. A song has a tonal structure, a harmonic structure, a rhythmic structure, and often words. All these elements may produce the same effect at the same time. A sense of melancholy, for example, may be created by a minor key, slow tempo, narrow pitch range, lower register, and, of course, the lyrics. Meanwhile, the music might slow down at the end of a phrase, telling listeners that a section is coming to an end. The pitches could do that on their own, but also doing it with the rhythm clarifies the macro structure of the piece.

Russo demonstrated the concept of redundant cues by playing a bit of "Hey Jude" (it was still the highlighted song in iTunes on the computer in the lab's control room). As the song built near the end of the fifth verse and the line is, "The movement you need is on your shoulder," he paused it and pointed out the tension The Beatles had created. There were notes that were not stable in the tonal hierarchy as well as a lack of stability in the rhythm. He hit play again. "So right here on 'Jude,' we return to the tonic, which is the most stable note in the key, and we also get a nice heavy kick drum happening there. So we know that we're stable because the rhythmic cues and the tonal cues are redundant."

Russo's argument that we don't hear music the way we

usually think we do certainly appealed to me. "We think our experience of music is about sound and it's about pitch, but we're wrong," he said. "Music is this mushy signal that is deeply moving, but we don't really know what it is, what it's trying to convey. That's maybe part of its beauty, trying to sort out what it is."

Before I left his lab the first time, Russo asked me if I had any plans to get back to my singing lessons. I hadn't been in a few months, mostly because I'd been too busy rather than because I was too dispirited. He had an idea for an experiment: he wanted me to continue my lessons and he'd track my progress to see if an amusic could eventually be cured. Well, that wasn't the way he put it, but that's what I wanted to hear. What he did say was if I practised diligently, he believed my brain could retrain itself and I could learn to sing and even score well enough to pass the amusia tests I kept failing. After all, stroke patients can learn to walk and talk again. Learning to sing shouldn't be as hard as that, should it?

Russo was excited about the idea. "It's a really unique opportunity for us," he said. "It's almost like you're making this up, it's almost too good to be true."

Later, I asked him why he'd wanted to do this experiment, one that seemed like quite the long shot. "The thing with you is you have the will—or some kind of crazy idea, however you want to describe it—that 'I'm going to do this,' that 'I'm going to do the music training,'" he said, adding that he suspected most people in my position would stay clear of anything to do with singing. "But you've jumped into the deep end." So it seemed like a great idea to assess me before I started and during my training.

I returned to the SMART Lab in March 2012 so Jonathan Wilbiks, a grad student, could get a baseline of my musical abilities before I restarted my lessons with Barnes. First, he had me fill out an Absorption in Music Scale questionnaire, which measures people's "ability and willingness to allow music to draw them into an emotional experience." I found out later that my responses put me in the top 20 percent of the population. "I'm not surprised," noted Russo when he emailed me the results, "but others might be." Alas, the notion that amusics can't enjoy music persists.

Then Wilbiks and I sat in a small room and started on the perception testing. I did the Montreal Battery. Although I was getting comfortable with what I was supposed to do, I wasn't necessarily getting any better at it. So much for the practice effects. Wilbiks also gave me the Beat Alignment Test (BAT) developed by John Iversen and Aniruddh Patel in 2008 at the Neurosciences Institute, which was then based in San Diego. Iversen is now at the University of California, San Diego's Swartz Center for Computational Neuroscience. Patel, one of the leading music cognition experts in the world, is at Tufts.

I didn't have to do all of the BAT, just the perceptual judgement portion, which seemed as though it should have been easy but wasn't. I listened to snippets of music — ranging from jazz to classic rock to stuff you'd expect to hear in an elevator — and a regular beeping sound at the same time. My task was to decide if the beeps were on the rhythm of the song or not. I also had to indicate how confident I was: guessing, somewhat sure, or completely certain. And Wilbiks timed me. When it was over, I admitted, "I don't think I did as well as I should have on that."

I returned ten days later to sing. The lab has a little recording studio at the back, and while two grad students worked the technology in the control room, Russo took me into the vocal recording area. "The first thing we're going to get you to do is sing 'Happy Birthday' from memory—and I know you just heard it this weekend," he said. I'd mentioned that I'd just returned from New Orleans, where my mother had taken the family for her eightieth and where the Preservation Hall Jazz Band had played "Happy Birthday" for her. "So you should be fine."

After I'd fumbled my way through that song twice, they asked me to do what Psyche Loui had made me do in Boston: listen to two tones and sing them back and then say if the second one was higher or lower than the first.

In an ideal world, I would have started this experiment before doing any sessions with Barnes. But Russo figured that since I hadn't had lessons—and hadn't practised—in months, I'd probably lost any advances I'd made in my first round of sessions. (Though Barnes would later tell me I hadn't regressed that much.)

The plan was to return to the lab every six weeks or so to repeat the tests. "I'd love to give you feedback along the way, but I do think it's going undermine the process," Russo said apologetically. "So I think your feedback is going to have to come from Micah."

"Wordy Rappinghood"

Dawson City, Yukon, is built on myth. By 1898, two years after the discovery of the first gold nuggets in a nearby creek, more than thirty thousand people made the mind-boggling trek to what they considered the Paris of the North. The most popular route meant hiking the Chilkoot Trail from Skagway, Alaska, to the Yukon River before travelling downriver to Dawson. More stampeders turned back or died than made it—and most of those who reached the Klondike discovered the easy-to-find gold had already been found. The exodus started in 1899 with another gold rush in Nome, Alaska, but Jack London and Robert Service stoked the Yukon legend with stories such as *The Call of the Wild* and poems such as "The Cremation of Sam McGee." And ever since, people have trekked to Dawson City to reinvent themselves.

In the last two decades, this town of 1,300 full-time residents—1,900 if you count the surrounding communities—has been pulling off its own reinvention by opening the Klondike Institute of Art and Culture, or KIAC, and then the Yukon School of Visual Arts. It also hosts annual music, film, and arts festivals. A quarter of the residents now work or volunteer in the culture sector. In the spring of 2012, I headed there—three flights, so not that treacherous—to be writer-in-residence at Berton House, the childhood home of Canadian popular historian Pierre Berton. I wasn't planning to reinvent myself as a singer, but I hoped that if I became more serious about practising, I could at least start to get a little bit better.

If you research Dawson before you go, you'll read about a bar called The Pit, characters such as Caveman Bill, and the Sourtoe Cocktail Club, which tourists join by drinking a shot of whisky with a human toe in it. Yes, it's true, the place is fun, but while I often patronized The Pit and once visited Caveman Bill's home in the west bank of the Yukon River—it's surprisingly comfortable—the Dawson I fell in love with is not the cartoonish one I'd read about. Still, Dodge, as some locals call it, is a funny place. I arrived at the end of March, as winter was loosening its grip, and though there wasn't much traffic, when I crossed the street, cars stopped at least thirty-five feet back. I found this endlessly amusing because back home in Toronto, pedestrians are lucky if drivers slow down before running them over.

Dawson's signature sight is the massive scar on the slope at the north end of town. Many generations ago, the Hän people saw a stretched hide drying in the sun and called it the Moosehide Slide, but when I look at this rare geological feature, I see The Rolling Stones' mouth logo. Which is fitting given how important music is here. The Dawson City Music Festival, which predates KIAC by two decades, has hosted many top Canadian acts and some international ones, including one of my faves, Bonnie "Prince" Billy. Not just the town's biggest event of the year, the festival begat a generation of community and arts organizers.

Within a few hours of arriving in Dawson, I found myself in The Pit, attending a fundraiser for a music scholarship. Operating since 1902, this drinking establishment is located in the Westminster Hotel, an example of the town's rich legacy of gold rush landmarks, buildings that are now in various states of repair. Dawson's setting—a flood plain along the eastern bank of the Yukon River—isn't classically beautiful, but the town is charming: dirt streets, wooden sidewalks, and pioneer architecture featuring colourful false-front facades.

As a band of local musicians did a short set, the first person I met asked if I played an instrument. Perhaps I should have taken her look of disappointment—or was it pity—as a warning because it soon seemed as though everyone in Dawson was going to get up on that stage. At one point, they Skyped in someone who had left town and a couple of people played with him. Two weeks later, I attended a coffeehouse concert at the art institute and, again, there was a steady stream of performers. At one point thirteen fiddlers and a pianist played together. Maybe this wasn't the right place for an amusic.

Fortunately, I am an amusic who loves music. So while I found it painful that I couldn't get up on a stage and perform, the Yukon offered me an opportunity to do something I've dreamed about doing since I was a kid: host a radio show. Starting when I was nine or ten, I listened to my bedside radio every night as I fell asleep. My parents' station was Toronto's CFRB, which was at 1010 on the dial; I listened to 1050 CHUM. And on clear nights in the winter, I could sometimes find a staticky WBZ 1030 in between those two. The Boston station broadcast the games of my beloved hockey team, the Bruins. Growing up with four sisters and no brothers put me in the middle of a weird dynamic: Although I was part of a large family, I sometimes felt like an only child. So music became my imaginary friend. When I went to my room to escape my sisters, the radio kept me company.

Located in a tiny blue building, Dawson's community station boasts a good-sized vinyl collection and room for not much more than what you need to broadcast. It was cozy. CFYT operates three or four days a week, depending on how many volunteers want to DJ. I took over the 3 to 4 p.m. slot on Saturday afternoons and dubbed my show *Face the Music.* Each week, I picked a theme: songs about place, songs about cars, songs about drinking, and so on. Although that wasn't an original concept for a radio show, I had fun with it.

What never dawned on me: Selecting songs based on the lyrics—rather than, say, the guitar riffs or the drum solos—and never playing any instrumental tracks might lend credence to the idea that I was just hearing the words. Isabelle Peretz's initial reaction when I told her how much I love music was to wonder if I was just attracted to the lyrics. And more than a few of my friends had joked about my fondness for literate songwriters and tolerance of idio-syncratic vocalists who spoke as much as sang. Inevitably, some people would grasp at my amusia as an explanation for my musical tastes.

I'd had two lessons between my baseline testing at the lab and my departure for the Yukon. At the first one, I explained Russo's experiment to Barnes. He was excited about it. "I love that you have a scientific counterpart to me," he said, "because I am the touchy-feely guy." He'd heard improvement over our first twelve sessions, but convincing me of that had not been easy, and I admitted that I never felt like I was getting better. Because part of what he does as a singing coach is help people get past their emotional or psychological barriers, having scientists track my progress was going to make his job easier. He was trying to turn me from somebody who had been told he can't sing—and believed it—into somebody who could improve. "When you really get down to it, your believing it is going to be the lube that makes it work. Without that, it's going to be a funky business for you," Barnes said. "I can drag someone kicking and screaming into improve-ment—I already have—but you're going to have to start leading the charge."

When I listened to the first dozen sessions again, I real-ized with a mix of shock and embarrassment just how nega-tive my attitude had been. I didn't believe I could do it. Not only was that hindering my progress, I could also under-stand how frustrating it must have been for Barnes. Part of the problem for me was that I had unrealistic expectations. I was always thinking that there would come a moment when it all would just make sense and I'd be able to hear pitches well and sing them back effortlessly. "It's like when you're doing math and all of a sudden you go, 'Oh, I get it, now I can know how to work this quadratic equation,'" I told him. "All of sudden, *click*." But I'd finally come to realize that this was going to be slow. And if the researchers I'd talked to were right, I may never even know if I'm singing in tune.

"That's disappointing," he said. "It'd be nice to go, 'I sounded great.'"

Still, he was hopeful that Russo's grad students and the experiment would help bolster my sense of improvement. "They get to be our little gold star achievement people in the lab coats. Love them. It's good because there's gotta be someone outside of this room who goes, Well done."

When he asked me if I'd been singing during the five months since our last session, I confessed that I hadn't. But he was understanding about that. He knows how little fun it is to practise something you're bad at. He also revealed the truth behind his nurturing approach as a coach. He offers some of his encouragement because that's what people need to hear to keep going. "That is true. There is also real improvement that I heard so I'm not making that part up. I might be exaggerating it," he said. "We've had some laughs about how it's not improving as quickly as we would both like it to be, so it's not like we're living in pretend world here."

And then, nearly twenty minutes after I'd arrived, we

started the lesson. We avoided the piano and he sat on the couch while I sat in a chair. We started with pitch-matching exercises and I was getting it right on the third try. That was consistent with what I'd been able to do in my first batch of lessons, though toward the end, I was getting it on the second try. "I was like, I don't understand how his ears work. They don't get it, they don't get it, and then they lock in," he said. "That's not somebody who can't hear. That's just somebody who takes a little longer to focus his hearing. His brain just takes a minute to get there."

But on this day, as we kept working, I kept improving. "*Mo mo mo mo,*" he called, and I responded, "*Mo mo mo mo.*" And I was getting better. "Third time . . . Second time . . . First time . . . First time . . . First time. You're hitting a million here."

We kept going. "Wow. You're doing great with these. It's like you've figured it out." I giggled. "Something's happening because you're nailing them."

"*Ma ma ma ma.*"

"*Ma ma ma ma.*"

"Can you tell you're getting them right?" he asked.

"Sometimes it sounds pretty good. Sometimes I think it's way off and you say it's right."

"If it's possible, you seem to have improved during this session," he said. "Maybe you don't need me."

Then he sang the first line of "Amazing Grace" and I tried to match him.

"Okay," he said with a laugh. "You need me."

One Saturday in Dawson, I devoted a show to songs about letters. Naturally, I opened with "The Letter" by The Box Tops, featuring a sixteen-year-old Alex Chilton on lead vocals. A lot of tunes about letters are full of woe and longing; apparently most good-news missives don't inspire songwriters. My playlist included "Take a Letter Maria," which was a huge hit for R. B. Greaves in 1969; "Please Read the Letter" by Robert Plant and Alison Krauss; and Mojave 3's "Return to Sender." To end on an upbeat note, the last track I played was Shuggie Otis's "Strawberry Letter 23."

I usually did a bit of research on the songs so I'd have something halfway intelligent to say as I introduced them. After reading a story — an untrue one, as I learned years later — about how Otis's girlfriend sent him letters on strawberry-scented paper, I googled the lyrics. Although I'd heard that song so many times over the years, and considered it a great R&B tune, I'd never taken in more than a few snatches of the words. I was surprised and delighted to see how lovely they are. The singer is such a happy guy and he wants this woman to know that receiving her letters, whatever they smell like, is such a treat.

When Carmen and I first started dating, music played a central role in our relationship. We often went out to clubs to dance, every house party worthy of the name ended with everyone dancing, and we attended a lot of concerts together. So when she visited me in Dawson, we did my radio show together. Predictably, we featured love songs. Once we'd agreed on our playlist, after some high-level negotiations, we realized many of our selections were about lost, unrequited, or damaged love: Elvis Costello's "Alison," Nick Lowe's version of "Poor Side of Town," "Love Hurts" by Gram Parsons. Twisted relationships make for the best love songs. But at least we were able to offer something for the true romantics with "Strawberry Letter 23." It was the only track that made it onto two of my shows.

Yes, knowing the lyrics to "Strawberry Letter 23" made me appreciate it even more, but I loved that song long before I learned them. Frankly, the suggestion that I'm only hearing the words when I listen to music has never made much sense to me. If that was all I cared about, I could just read poetry (which I'm ashamed to admit I don't do as often as I should) or listen to audio books (which I never do) and podcasts (one of the few I listen to regularly is NPR's *All Songs Considered*, a podcast about music). Others may have had their doubts, but I was convinced my response to music had to be more complicated than that.

Before I flew up to the Yukon, I bought an iPod Nano and loaded my last two sessions with Barnes on it. (I wanted to keep them separate from all the songs on my iPod Classic, which I usually listen to on shuffle. I'd be horrified if one of my lessons came on.) I added a couple of songs we had been working on, including Johnny Cash's "I Walk the Line." Then every day in Dawson, I would practise while doing stuff around the house: tidying up, washing dishes, and so on. I also created a spreadsheet to track how much time I devoted to this project. I did about forty minutes a day. It was quite painless because I had so few responsibilities that I never had the excuse of being too busy that life in the city gave me.

For most singers, working with an iPod would not be the ideal practice method, because with the ear buds in place, they wouldn't be able to hear themselves properly. But I couldn't hear whether I was hitting the notes. And the iPod kept me much more focused. When I'd practised to recordings of my lessons on computer, it was too hard to not get distracted. I couldn't resist checking my email and or finding other things to do on the Internet. Still, it bothered me that I didn't know if I was hitting the pitches I was trying to match, and I worried that I was just reinforcing bad habits. I belted it out in Berton House anyway. At the beginning, at least, I didn't need to worry about other people hearing me: it was still too cold to open the windows.

I wasn't ready for anyone to hear me. I entered the Dawson City International Short Film Festival's 1 Minute Film Contest and attended the screening of the entries, followed by karaoke, at Diamond Tooth Gerties Gambling Hall (once Canada's only legal casino). New friends who knew I was taking singing lessons urged me to get up on stage. But that certainly wasn't going to happen.

I'd already had three bad karaoke experiences in my life. One night, back in the early 1990s, Carmen, who was not yet my wife, and I found ourselves leaving one of her work parties and piling into the back seat of a car on its way to a Korean restaurant to do something called karaoke. The do-it-yourself singing craze was a fairly new phenomenon in Toronto, so I had no idea what I was in for. The videos all seemed to feature doleful young Korean women clutching books to their chests as they walked home from school, but our friends sang a few songs and soon we had no choice but to try it. Carmen picked a song we both loved: a mid-'80s gem, the Fine Young Cannibals' cover of Elvis Presley's "Suspicious Minds." We weren't paying attention when our turn came up. By the time we rose from our table and made our way to the mic, the lyrics were already crawling across the screen. We scrambled to catch up but before long, I heard the woman at a nearby table groan, "Ah, you suck."

I resolved never to do that again. And for a long time, I didn't. But, eventually, at an end-of-season hockey banquet, my friend Steve Watt insisted I take advantage of the karaoke machine. "We'll do a duet," he said. Late in the evening, and well primed, we took the stage to perform a rendition of The Band's "Up on Cripple Creek." A few notes in and I was in a complete flop sweat, refusing to lift the microphone higher than my crotch.

The last time was, thankfully, at one of those places with private rooms, getting near closing time. I screamed a lot. Nobody complained, presumably because we were all too drunk to care.

But if anything, I had become even more self-conscious about my musical ability since starting my singing lessons and being diagnosed as amusic because of my increased awareness of how bad I was. I wasn't just bad, I was scientifically bad.

Fortunately, Carmen remained forgiving. Before my stint in the Yukon, I'd taken her to Spain for a big birthday. One night we were walking around Barcelona and came upon a man wearing headphones and singing along to whatever he was listening to. "See?" Carmen said. "At least you're better than that guy."

After a month in Dawson City, I returned to Toronto for three days to attend the launch party for a book I'd helped write. I squeezed in a trip to Russo's lab to do the interval matching and sing "Happy Birthday" a few times. Then I had a lesson with Barnes. As usual, we started by chatting for a bit—not just to catch up but so he could get a sense of my attitude. I once asked him if he talked to his other clients as much as he gabbed with me and he said that, with some, even more.

Eventually, he said, "Let's see where you're at."

"*Ma*," he sang.

"*Ma*," I responded.

We did this for a few minutes until he said, "You're on pitch. That's the best you've ever been."

We did more exercises. "Wow," he said. "Not much to do in the Yukon. You're actually practising." He was right about the practising but definitely wrong about things to do in the Yukon. "This is a marked improvement. I've never heard you this good."

But when he asked me to match two notes, I didn't do as well. At least initially. But we kept going until he said, "Your ears are so much better that it's freaking me out."

Then we moved over to the piano and I tried singing "I Walk the Line." Barnes's review: it was more like talking and less like singing, but I was getting close to being able to sing it at karaoke. I reminded him that singing a song is a big leap from matching notes. Besides, Cash talk-sings, too.

"He's the perfect singer for you," Barnes agreed. In fact, he said I was ready to work on more songs, not just pitch matching, and suggested I find another Cash song to learn. We listened to "Ring of Fire." Barnes said it was harder than "I Walk the Line," because the phrasing is more difficult, but not so hard that I shouldn't give it a try.

Barnes suggested that because Russo — an actual scientist — figured I could improve, I was starting to believe it myself. My singing was louder and more confident. "Your attitude is different," he said. "You're hungrier for this."

That was a reassuring message to hear before heading back up north.

During one of my lessons with Barnes, after I'd been working on "A Good Year for the Roses," I admitted that I'd been surprised to learn that the line was "the lawn could stand another mowin'" instead of "no one could stand another moment." Inevitably, this led to a discussion about misheard lyrics. Barnes told me about the time when he was a teenager that he bet his brother twenty dollars over whether or not Aretha Franklin was singing, "sick of your fucking ways" in "Ramblin'." Barnes insisted, "Daniel, she's not singing, 'sick of your fucking ways' — she goes to church." He was right, of course, but never collected on that wager.

In the summer of 1969, the year I turned eleven, my sisters and I discovered a copy of the *Hair* soundtrack at the Muskoka cottage my parents rented every August. We played the hell out of it. Covers of some of the tunes were already radio hits ("Aquarius/Let the Sunshine In" by The 5th Dimension, for example, was the number one song on CHUM's Top 100 of 1969). But I soon learned the others and started singing "masturbation can be fun," a line from "Sodomy," everywhere I went.

The au pair my mother had hired for the summer pulled me aside. "Do you know what that word means?"

"No," I admitted.

"Well," she suggested—wisely, I later realized, "I don't think you should sing it until you do."

Although I found my conversation with this teenager more puzzling than awkward—at least until my vocabulary improved—it reinforced the message that maybe I wasn't meant to be a singer.

Misheard or misinterpreted lyrics have always been with us. There's even a word for the phenomenon: mondegreen. One of the most cited examples is "'scuse me while I kiss this guy" instead of "'scuse me while I kiss the sky" from Jimi Hendrix's "Purple Haze." Misheard lyrics can be fun, but so can misunderstood ones. When I first listened to "Your Little Hoodrat Friend" by The Hold Steady, I thought of Bruce Springsteen singing about young women without shoes sitting on the hoods of automobiles despite soft precipitation in the summertime. But those were '70s hoodrats; today's hoodrats aren't so innocent. Meanwhile, some people completely misunderstand the point of songs. American politicians have played Springsteen's "Born in the U.S.A." proudly at campaign events, apparently unaware that the lyrics are an indictment of the way the government treated veterans of the Vietnam War. Part of the problem may be that its anthemic sound is at odds with the lyrics.

Many songs will naturally have words and music that match emotionally — a mournful song will have sad lyrics, for example, while a happy one will have upbeat lyrics. But sometimes there's a surprising, or even creepy, mismatch between the words and the emotion of the tune. Some bands — Belle and Sebastian, for example — seem to specialize in this. Even better, many songs have ambiguous lyrics that allow listeners to imagine their own meanings.

Sometimes it's obvious why we like certain lyrics and sometimes it's not, which can make the appeal of the words another mystery of music. When we're heartbroken, every love song on the radio will somehow capture exactly how we feel. With political songs, we might agree with the singer's position on an issue and that makes us fall for it. Would we love Sam Cooke's "A Change Is Gonna Come" even if the words were nonsense instead of a powerful message about racism in America? Maybe we would because the arrangement is so lush and beautiful, though perhaps not as much. And I don't think we can say even that about most protest songs.

Sometimes it's the storytelling: "Acadian Driftwood," for example, which recounts (with some poetic licence on Robbie Robertson's part) the expulsion of francophones from what's now Nova Scotia and their move to Louisiana, is easily a Top 5 track in The Band's canon. We also fall for songs about places, both ones we've been to and ones we'd like to go to. One of my favourite Ron Sexsmith numbers is "Lebanon, Tennessee," which he wrote while working as a courier. One day he saw a package with the city's post-mark on it. Perhaps that's a case of the story behind the lyrics making them even more compelling, but it is still a gorgeous song.

Sometimes it's just one line that comes to mean so much to us (even if it's out of context). And often the singer's delivery is more powerful than the meaning of the words. In fact, the lyrics don't even have to be clever for us to like them. Party songs are ones we enjoy belting out with friends, usually bolstered by booze, and the rousing music seems to give the words extra meaning. And then there's "Louie Louie," the rock 'n' roll standard that no one knows the lyrics to but everyone loves.

My attraction to lyrics occasionally baffles me. I have no idea what the Calexico song called "Sunken Waltz" is about, but I adore the sound (an intoxicating mix of twang and mariachi) and there's one line in it— "Tossed a Susan B. over my shoulder"—that just slays me. I don't really know why the singer is throwing away a now-discontinued silver dollar like that, though I assume it's for luck because the next line is "and prayed it would rain and rain." But hear-ing those words always makes me happy, even though the song conjures a sense of desolation. Words: so weird, so wonderful. Just like music.

Yes, the words mean a lot to me. They may even help me fall in love with a song, but they aren't what first attracts to me to it. And I'd likely be keen about lyrics even if I didn't have a pitch discrimination disorder. After all, a fascination with words is something of an occupational hazard when you're a writer. So it shouldn't be a surprise that I have a fondness for literate songwriters such as John Darnielle of The Mountain Goats, Craig Finn of The Hold Steady, and John K. Samson of the now-disbanded Weakerthans.

But much of the music I love doesn't slot into that category. My favourite album of the last few years is *Lost in the Dream* by The War on Drugs. I've listened to it so many times and yet I couldn't recite one line from it unless I put it on and heard it first. That album is not really about the words; it's about the dreamy mood the band creates. The sound is fresh and inventive, yet owes so much to influences from the past, from Roxy Music to Bruce Springsteen to Kraftwerk and lots in between. I also like plenty of music with banal lyrics. Or even with nonsense lyrics as in "I Zimbra," the Talking Heads' song from *Fear of Music* that combines the dadaist poetry of Hugo Ball with African rhythms. Rare is the song that I hate just because of the words, though "People Are People" by the otherwise enjoyable Depeche Mode is an example. (Aside from the vapidity of the lyrics, surely rhyming "be" and "awfully" warrants prosecution.) And with some music I love—early R.E.M., for example—who knows what the singer is saying?

In a few genres, notably folk music, the words are the focus and we are likely to catch at least some of them on a first listen. With other genres, we may take in a few lines initially—the obvious hook in the chorus of a pop song, for example—but it usually takes at least a few more spins to really catch the words. I once created a mix CD for some friends going on a fishing trip. When they returned they said they liked it, but they'd been listening to it and thinking, *Wow, a lot of these songs are about death — is he trying to tell us something?* Sorry, guys, I assure you it wasn't intentional.

I don't think I even hear the words that well when I listen to music, especially on the first few listens. Despite many protestations from people who say they are interested in some other dimension of the music — the melody, for example — they all seem to hear the words better than I do. And anyway, just about everyone ends up being pretty deeply into the lyrics. Even rock critics aren't immune to this. Read album reviews and you'll find that many writers pay way more attention to the lyrics than the sound, which they usually slot into a genre or combination of genres.

This focus on the words may simply be because writing about music is like dancing about architecture, to borrow actor Martin Mull's famous aphorism. In March 2015, Jon Pareles of the *New York Times* wrote a piece called "Courtney Barnett Prepares Her Debut Album" for the paper. Barnett had already become a favourite with some indie rock fans after a release called *The Double EP: A Sea of Split Peas,* and I'd seen her at Toronto's Lee's Palace in the fall of 2014. But her new offering, called *Sometimes I Sit and Think, and Sometimes I Just Sit,* was getting lots of advanced hype, including from the *Times,* which isn't exactly known as a trendsetting publication among music fans. (Though I do find the paper's writing on the subject generally strong.)

Pareles's piece was almost entirely about Barnett's lyrics. "The Australian songwriter Courtney Barnett writes about the little, not always simple, things in life," it began. "An asthma attack. Insomnia in hotel rooms. House-hunting on a meagre budget. Trying (and failing) to impress the swimmer in the next lane at the pool. Roadkill." While he talks at length about her lyrics, his references to her music are mostly in passing: "a hurtling, distorted, four-chord rocker"; "the through-line from grunge back through garage-punk to the mid-1960s"; and "smart, scrappy guitar-based songs." He has more to say about her singing, though, noting that it "moves from the conversational to the urgent without coming anywhere near pitch correction; when her voice droops or scratches, it sounds only more candid."

Ah, the voice. A singer's voice has the potential to be far more powerful than the words that voice is singing. I'm convinced that I respond to it more than the lyrics. Vocal music is easily the most popular type of music. Opera and a few acts such as Iceland's Sigur Rós and Inuit throat singer Tanya Tagaq aside, we generally prefer to listen to people singing in a language we can understand. Gary Marcus writes that vocal music is more popular because it offers the immediacy of words and the expressiveness of the human voice, but it also helps add repetition and variation.

Ethnomusicologist Gillian Turnbull believes that when we listen to vocal music, we react to three powerful elements. The words are part of it: they are a form of poetry, and they often tell a story or express strong emotions. But there's also the sound of the voice, which we will react to before we even understand the lyrics. The voice is one of the most basic forms of human expression. When we speak, our intonation, timbre, and register can convey even more meaning than our words. It's also a powerful identifying feature of a person: the voice tells us about the singer the way the bouquet tells us about a wine: top notes of age, gender, experience, emotion, and illness with strong hints of personality. In music, these qualities interact with melody, pitch, and rhythm to make singing even more powerful. The third element is the music itself, which generates its own emotional reaction. "We have a visceral, physical reaction to music, and that's without even considering the role of the voice," said Turnbull. Put all those things together "and suddenly the singing voice is probably the most expressive thing that you could find." No wonder it's so important to us.

For several years, Elvis Costello's albums practically served as the soundtrack to my emotional life, even as he experimented with different styles. *Imperial Bedroom* came out in 1982, the same summer I owned my first Walkman. That fall, I went to school in Ottawa for a year. The walk from my basement apartment in Centretown to campus was about forty-five minutes, the length of one side of a C90 cassette tape and almost as long as Costello's album. So the logical thing to do was to try to learn the lyrics to every song during my commute. I did pretty well at this project, which says a lot about how much I adored that record.

But I first fell in love with Costello's music when I played *My Aim Is True* obsessively while suffering through a Montreal funk. It wasn't his verbal gymnastics, which became even more acrobatic in subsequent albums—it was his angsty rage. And the anger of the lyrics was only part of it. So I'm sure my reaction to the music I love says a lot more about something other than the words.

"Small Change"

Erase the tapes. That was Ted Nugent's ten-million-
dollar plan. Along with his song "Cat Scratch Fever" and
his outspoken opposition to gun control, the Detroit rocker may be best known for his attempt to buy Muzak, the
company that produced elevator music, for ten mil in 1986.
Even people who aren't fans of the self-styled Motor City
Madman secretly cheered the idea, because to most of us,
corporate aural wallpaper is a joke. Or at least nothing we
would ever want to listen to. But there was plenty of science
to the music Nugent wanted to destroy.

Say what you will about Muzak, but the life of George
Owen Squier, the man who started it all, sure wasn't vapid.
He went to West Point despite only a grade eight education; achieved the army's first doctorate when he completed
a Ph.D. in electrical science at Johns Hopkins University;
patented many innovations, including multiplex telephony,
which allowed the transmission of several calls over the

same wires at the same time; rode in an early test flight with Orville Wright—making him the first-ever air passenger—and devised the specifications for the first military plane; served as the chief signal officer for the U.S. Army, organizing intercontinental radio and cable communications during the First World War; and created the army's first air corps, precursor to the U.S. Air Force. He left the military in 1922, but didn't retire. Instead, he pitched some of his radio patents to the North American Company, a Cleveland-based utility. It created a subsidiary called Wired Radio, which pumped music to subscribers in homes and stores over wires. Before he died in 1934, Squier rebranded the business as Muzak, a name he came up with by combining music and Kodak.

As wireless radio, which was free because it was supported by advertising, became increasingly popular in homes, the company concentrated on commercial clients. In the 1940s, factories were enticed by the Muzak research department's claims that the service would make workers happier and more productive. Under the category Stimulus Progression, for example, Muzak organized its programming to counter employee fatigue by moving from subdued songs to more invigorating ones in fifteen-minute chunks. According to *Elevator Music: A Surreal History of Muzak, Easy-Listening, and Other Moodsong* by Joseph Lanza, "programs were soon tailored to workers' mood swings and peak periods as measured on a Muzak mood-rating scale ranging from 'Gloomy—minus three' to 'Ecstatic—plus eight.'" The good times with the bad music didn't last forever, though. The company filed for bankruptcy protection, and in 2011 Toronto-based Mood Media bought it so businesses could continue to get their music fix.

Muzak fascinates Frank Russo, a fact I learned after it struck me that some record producers might be intrigued by Russo's work. "It was a real think tank of researchers trying to understand how changes in the music lead to changes in mood to influence worker productivity, to influence buying decisions," he said. Unfortunately, the company has always kept its research proprietary, refusing to publish it in academic journals or otherwise share it with people like Russo.

But scientists have done their own work on what music does to us. The Mozart Effect, for example, is an idea that emerged after researchers tested the spatial abilities of undergraduates who had listened to ten minutes of classical music, relaxation instructions, or silence. The students performed better after listening to the music, which happened to be Mozart. Later, other researchers discovered that any pleasurable music—even pop—has the same effect, and the temporary positive benefits are not confined to spatial abilities but extend to other intellectual abilities relating to pattern recognition and memory. But that doesn't mean that all music works the same way (sad pieces, for instance, don't), and other activities that are pleasurable and stimulating (such as listening to a story) could have the same effect.

Most knowledge, it seems to me, can be used for good or evil. That's probably true even when it's about something as great as music. So I asked Russo if it would bother him if a record producer cynically made a pop song using all the research that he and his music cognition colleagues had created. I doubted that it would be the kind of music I would like, but I long ago realized that what I like and what's a hit aren't necessarily the same thing.

His answer surprised me. "I would endorse it," he said. "I would love it." He admitted that the music would probably be horrible because while researchers are slowly learning cool things about music, there's still plenty they don't know or understand. And in the same way that it's hard to build a robot that speaks in a human way, we're a long way from having a robot produce songs. "There's a kind of magic to music making that is hard to manufacture. It is an intensely human kind of activity," Russo said. "But I'd buy it."

One song I would never buy is "Happy Birthday." Of course, there would be no reason to ever want to own it because we hear it all the time. Still, it's a song just about everyone seems to sing badly. It's not particularly easy, and if several people belt it out together, they'll all sing it in a different key and the bad singers will inevitably overwhelm the good ones. (Barnes once joked that he invites only good singers to his birthday celebrations.) And yet "Happy Birthday" is the song music labs make subjects in their experiments sing because it's something everyone knows regardless of age or culture. I never particularly liked that song—the lyrics are beyond banal—but having to sing it in the SMART Lab really made me hate it, especially since I didn't have a party full of people to drown me out. I wished that the lab could have just tracked my progress on, say, "I Walk the Line" or another song I was working on with Barnes.

Each time I went in for another round of testing, I stepped into the recording room and stood behind the microphone. I would sing "Happy Birthday" twice, then do the interval matching task, and finally, sing "Happy Birthday" two more times. A couple of Russo's grad students sat in the sound booth on the other side of the glass. They pulled down a screen so I wouldn't be distracted by them, but I could still see enough to make out when they were talking and laughing, which was disconcerting. Were they talking and laughing about my singing or about what they did on the weekend?

I tried not to think about that, but I decided that if I had to sing that friggin' song, I was going to try to get better at it, even if practising on the side was cheating. (I later discovered the lab didn't mind at all.) I searched online for a version I could work to and, much to my surprise, had no luck. That's when I discovered the song —properly titled "Happy Birthday to You"—was not in the public domain. In 1893, sisters Patty and Mildred Hill composed "Good Morning to All," which eventually morphed into the little ditty no one can avoid now. Though it appeared in a songbook in 1922 with the "Happy Birthday" lyrics, no one copyrighted it until 1935. And due to changes in copyright law since then, the protection could have lasted as long as ninety-five years.

Although you wouldn't be dinged for royalties if you serenaded a friend over a birthday cake, you had to cough up some cash if the performance was in any way commercial.

But because of the song's complicated history, the legitimacy of the copyright was a matter of some debate. And in 2013, Good Morning to You Productions, a company that had paid $1,500 to use "Happy Birthday" in a documentary film, launched a class action suit against Warner/Chappell Music, which held the copyright and collected about two million dollars a year in fees from the song. Or did until the fall of 2015, when a district court judge in Los Angeles ruled the copyright wasn't valid. If a higher court overturns the decision, we will have to wait until 2030 to find "Happy Birthday" on YouTube (or whatever 2030's version of YouTube is). But if the ruling stands, we might have to hear the song even more often.

At my next lesson I asked Barnes if he'd help me with it. He said sure and asked me to sing it. "That's good," he said when I finished. "It's just out of tune. But you understand the melody."

We moved over to the piano to see if that would help with the tuning, but I was much worse. So we worked on it a bit before doing some pitch matching. When I did well with that, he asked me if I could tell that I was doing well. "Not really," I said, "but I feel more confident because I've been practising."

Then we moved on to "I Walk the Line." I sang along to Johnny Cash, but Barnes hit pause on his computer partway through. "You're not quite where I want you. You're not singing his melody — you're singing his melody in a weird harmony."

I tried again. "Now you're in pitch with him — you found it," he said. "Nice."

I kept at it and after I sang the chorus a bit weirdly, he said, "Sure, it's a harmony. I'll go with that." And then he laughed. What he meant was that I was singing all the wrong notes, but I was consistently off.

I did it again. "Intermittently," he said, "you're dead on."

Faint praise was better than no praise, I guess.

In a later rendition, I really sang my heart out on one line and it came out quite wonky. "Go for it," he said and then laughed and laughed. "Who cares if these are the right notes, I feel this." He kept on laughing.

This was my first lesson since returning from the Yukon. Although I hadn't seen Barnes for a few months, it had gone well. He asked me to work on "Happy Birthday," "Amazing Grace," and "I Walk the Line" for the next session. When I returned, he asked me how my singing was going, and I admitted that I was frustrated. "With the songs, I still don't feel like I'm making progress," I said. "I still feel that even if I can match tones, I can't sing a song."

"I'm used to people having all sorts of feelings and reactions to where they're at," he said. "Disappointments. Expectations. That's kind of what I deal with all day."

"I understand that. I'm just letting you know where mine are," I said. "I'm sure most of the people you deal with are having frustrations that are a lot loftier than mine."

"I see them all as equal, actually. I really do. Somebody who has a recording contract and a tour that's booked probably feels as frustrated with not getting their breathing together as someone like you who can't tell relative pitch properly and doesn't feel an improvement. I think those are comparable feelings of frustration."

One problem with this process was that I was always working to a recording of my last session. So if my last session had not gone well, I was reminded of it every day as I practised, reinforcing my frustration.

Barnes was sympathetic — to a point. "There are times when you're not even close and there are times when you're dead on, though. Your reactions to it are very strong, but you can't sing, so what are you expecting to happen?"

"Progress..."

"I hear the progress," he said. "You expect yourself to be able to sing songs better than you can right now."

We did some pitch matching exercises. "Feel how you're nailing it?" he said. "Can you tell?"

"On the fourth try," I responded, unimpressed.

"Okay, your attitude about not being able to do this and how bad you are is not going to help you," he said. "Lose it."

I'm sure he was as frustrated with me as I was with my singing. He'd briefly let it show, but later he was back to his usual upbeat approach. I laughed and said, "It's funny how you can turn anything into a positive."

"It's my job, because you can turn anything into a negative," he said.

Still, he made it clear that I'd regressed physically. I wasn't breathing as well and I was tighter because I was too focused on getting the notes right. My bad breathing was hurting my ability to hit the notes. He didn't want me to leave with negative vibes, so before I took off, he assured me, "You have nothing to be depressed about. I think you're coming along."

He started the next session by saying, "Last time we

met was a challenge for you." But on this day, I was feeling better. I'd been working on "Happy Birthday" and was more confident about the shape of the song, the ups and downs of it.

After pitch matching and more work on breathing and supporting the notes with my air, we started in on "Happy Birthday." He had me just talk it a couple of times and then sing it.

"Happy Birthday to—"

"It's a little higher," he interrupted.

"To."

"Little lower," he said and then sang "to."

I tried again: "to." But a lot of times when he asked me to adjust a note up or down, I sang back the same note.

"Little lower," he said again before singing "to."

"To."

"Little lower... to."

"To."

"Little higher... to"

"To."

Finally. He sang his response this time: "That's the note."

I sang the whole line: "Happy Birthday to you."

"Best yet," he declared.

We did it again.

"You're overshooting your 'to'—tiny leap."

And that's the way we worked that song. At the end of the hour, we both felt pretty good about how the session had gone.

In late October, not even eight months after I started the SMART Lab experiment, Jonathan Wilbiks sent me an email with a subject line that read: "Next session and debriefing." He wondered if I could come in the following Tuesday. "This would be for a final administration of the battery of perception tests you did at our first meeting. It would also give us an opportunity to share with you some of the results that have come out of the four recording sessions." Final? Results? The experiment was over? Had it been such a miserable failure that they were pulling the plug on it already?

I went to the lab and sat down with Wilbiks in one of the booths. As he explained that he wanted me to listen to two melodies and indicate if they were the same or different, I said, "You know I've done the Montreal Battery about a million times?"

"Yeah, we know."

"But you want to see if I'm getting better."

"Pretty much. Depending on who you talk to, it may or may not be possible to get better, but we hope it might be possible."

I found out later that it wasn't possible, at least for me, to get better on that infernal test. Practice effects, my ass. But I did improve slightly in the Waltz or March segment: I'd scored 22 the second time, up from 18, though the average person scores 26. Besides, it was hard to say if that was just a fluke. My results on the Beat Alignment Test the second time around were more encouraging: I'd been right 83.3 percent of the time, up from 66.6 percent, and not too far off the average of 90 percent.

By the time Wilbiks and I were finished, more than an hour later, Russo had arrived. I was nervous about what he would say as the three of us sat down in the lab's main room, because it seemed way too early to give up on the experiment. And, at first, Russo played coy with me. "You probably have some idea how things are going already. So why don't we start with that: What are your thoughts?"

"It's a little hard to know, because Micah's very supportive, encouraging," I said. And it was true. I was never entirely sure when Barnes was being honest in his praise and when he was just being nurturing. "I don't think—though you may tell me something different—but I don't think there's any doubt that I am better than I was. There are definitely some frustrations, times when I think I am not making any progress, but then there will be some progress."

After hearing that, Russo told me about the results of the experiment. It did not start out well. He and his grad students had seen no improvement in my ability to sing the pitch pairs. That didn't mean that Barnes was fibbing to me when he'd tell me that my ability to match his pitches was improving. Even for non-amusics, it's harder to match instrumental or electronic tones than it is to match the human voice. Most of the time when I did ear training in my lessons I'd used Barnes's voice as my model, but the lab used piano tones in the experiment. Barnes had tried getting me to match piano notes a few times, but I was so hopeless at it that he invariably soon abandoned the idea. Still, I asked Russo, if I worked on matching isolated non-human pitches, would I get better? "I think definitely," he said.

But he and his grad students hadn't only been looking to see if my ability to match tones would improve. Russo didn't go into it then, but later he gave me more details about the pitch pair part of the experiment. It included both a production task and a perception task: I had to sing the tones back, and I had to say if the second one was higher or lower. They wanted to find out if there was a connection between my ability to reproduce that pitch interval — sing them back high to low or low to high, regardless of whether or not I managed to nail the notes — and my ability to perceive whether the pitch was going up or down. Or would there be what researchers call a dissociation.

If we think about pitch processing as one big package, the natural assumption would be that there's a correlation between the two skills. But that's often not the case with amusics, who may be able reproduce the interval — even if they can't hear it accurately. "So you might have sung back a low to a high tone to me, but then reported that, 'Oh, pretty sure it went down,'" explained Russo. "And you'd be fairly confident about it." This dissociation fascinated him, even though he wasn't the first researcher to find it in an amusic. He believes it reveals something about the way brains like mine work. "In short, it tells us that there's more than one route for pitch processing."

But he did have some good news for me. The other part of the experiment had been my singing of that dreaded song. "We actually don't see any improvement in the isolated pitches or the intervals, but we see a massive improvement in 'Happy Birthday.' It's a huge improvement." In the last performance, my pitch error rate had dropped to an average of five cents (100 cents is a semitone, or one-twelfth of an octave). "Five cents is better than most people can sing 'Happy Birthday,'" he said. "So how in the world are you doing this?"

These results sounded too good to be true. And they were.
Russo later explained that a calculation error had made my progress seem much better than it was. My best performance—at the last recording—had been 92 cents off. That didn't sound impressive at all to me, though it was better than my worst, which was 132 cents off. "The big picture is still as it was (you had some improvement) but less impressive," he emailed. "I would say you might be in left field, but you are definitely 'in the park.'" Still, it wasn't all bad news. The lab had used computer algorithms to analyze not only my pitch but also the contour. And I was quite good—better than most people—at getting the contour right. "There are almost zero contour errors! When the melody goes up, you go up...However, this aspect of your singing was strong from the onset of testing."

Knowing that music is about much more than pitch, Russo had wanted to look beyond that dimension. To do so required some qualitative assessments of attributes such as timing, phrasing, and tone quality, and that's why he invited about fifteen people associated with the lab to a listening party. They sat around and evaluated randomized versions of me singing the song, scoring them from one to ten. On this subjective, but still useful, basis, my last renditions were definitely better than my earlier ones. Not perfect, of course, because I wasn't hitting all the notes. But along with the strong contour, the timing was excellent.

Russo was intrigued that he'd seen such improvement in my singing despite my difficulty matching pitches. "You're not improving on the production of isolated tones and intervals, but you are improving on the production of a full melody," he said, adding sensibly, "which is all anyone would care about in the real world." He'd expected both—or neither—to improve. Wilbiks added, "It would have made more sense."

Had I improved by listening to the song again and again and trying to sing it, Russo wondered, or had I only improved because Barnes had repeatedly corrected me?

I took a big breath. "Okay, here's my theory: I learn the contour—the shape—of the song." So once I discovered that the "to" in the second line of the song, for example, is a little higher, I'd try to sing it that way. But then Barnes would often tell me, "No, you're going too high, it's just a little bit higher." And then I would take another stab at it. "In my mind, that's the way I've learned it. Then the problem is I know it's higher here, but I don't know how much higher."

"So how do you manage that?" he asked. "How do you figure that out?"

"I don't know. Maybe just practice."

"But somehow it would seem that you are listening to your own voice and matching it to Micah's voice and saying that's different, let me try again."

He wondered if there might be a difference between

well-encoded melodies in my long-term memory and little fragments I've just heard. Maybe it was possible for me to have better pitch production when I know a piece really well after hearing it hundreds of times. He proposed that a possible follow-up experiment would be to learn one song with Barnes and a different song on my own just by listening to it a lot. "When I learn a song, that's what I do," he said. "You're having some problems with pitch matching, so maybe you do it differently. I guess I'm trying to figure out how is it that you're doing this."

I'm afraid I wasn't much help with this mystery. After all, I still could never be sure when I was singing the right notes and when I wasn't. "I just assume every one of them is wrong. But the thing about 'Happy Birthday' is that in my mind I have an idea of the shape of the song, but I still don't have that confidence, like, Yeah, I nailed it." I compared the experience to when I used to play golf and could never develop a repeatable swing, one I could rely on every time I addressed the ball. That's why I was an erratic golfer. I made a lot of bad shots mixed in with some good ones, so I wasn't able to break 100 often enough to really love the game. "That's the way singing is for me at this point."

In the end, he never did ask me to try to learn a song on my own. In 2013, we had lunch at the University of Toronto's Massey College, where he was a fellow while on sabbatical from Ryerson. Russo admitted that he has a tendency to take on too many projects so many of them don't end up going anywhere, but he still writes five to six peer-reviewed papers a year as well as conference papers. I was relieved that we didn't do that experiment, because I'm pretty sure trying to teach myself a song would be a disaster. After all, my attempts to sing a song I knew well for the first time usually prompted Barnes to say I knew the rhythm, but the notes were all wrong.

Russo had more questions than answers, but he thought there were a few possibilities. Singing requires auditory-motor planning: our brain has to make a plan for the sequence of necessary muscle movements. Then those commands have to go to your muscles, which have to contract in the right ways at the right times. Most people would be able to accurately perceive what they're singing and compare it against some memory representation of what it should sound like, making any required adjustments to the motor plan. "It's also possible that perception is not just auditory, it's also some kinesthetic feedback."

He was referring to the sensation of the movement of muscles and other body parts. But it's also a way of learning. Kinesthetic learners tend to develop skills by doing an activity and relying on trial and error to get better. According to one popular model, there are four learning modes: watching (reflective observation), feeling (concrete experience), thinking (abstract conceptualization), and doing (active experimentation). Many people use more than one of these techniques.

Learning to sing isn't just a matter of developing auditory motor control; part of it is having an awareness of what your body is doing and developing muscle memory. Maybe I'd been able to develop muscle memory for a song after I'd spent a lot of time on it. "And maybe you're able to use that. I think we all use that to some extent, but we tend to focus on auditory," said Russo. "Maybe out of necessity, you focus on that."

That seemed possible. I couldn't tell when I was hitting my pitches, but I could often tell when I wasn't because it just didn't feel right.

I also pointed out that my approach to singing is not intuitive, as I assumed it was with most people, but something more deliberate. "For me, it's not so much a motor plan, it's an intellectual plan."

That was another possibility. Maybe I could go straight from that memory to the motor plan as long as that memory is really crystallized. "If that were the case, I don't think you could ever be a great singer because you can't fine-tune," he noted, in what was surely a statement of the obvious. "But maybe there's some means of simply going from well-rehearsed pitch memories straight to pitch production." Somehow, in the feedback loop in my brain, I was bypassing the pitch-hearing part.

If practising one song had helped, I wondered, could I also improve my ability to discriminate between different pitches and sing them back? And if I did, would that help me to really learn to sing? After all, my ability to do an okay job on one song seemed more like a weird party trick than really learning to sing.

Someday, if he gets the time, Russo might use the data from the experiment to publish a case study about congenital amusia and the possibility of retraining the brain's circuits. "The story will be a complicated one because it's not clear cut," he said. "It's not that you spent eight months training and now you're ready to go to Carnegie

Hall. You've spent eight months training and some aspects of your music production have improved and, in particular, the pitch processing, which is the core of tone deafness, has had limited improvements."

While he still had a lot of questions, I still had the one I'd started with: Did these results help him understand—or change his ideas about—what he thought I was hearing when I listen to music?

"I assume that it's not about the harmonies," he started. Beautiful harmonies and singers moving away from the centre of a key and coming back to it probably wouldn't mean much to me. While I have no doubt I'm not hearing harmonies the way I'm supposed to—the blending of different voices appeals to me, but I usually can't tell the difference between unison singing and harmony—they still sound good to me. When I told him that I like the harmonies of The Beach Boys, a band that I cited only because it was the first example that came to mind, not because I listen to them often, Russo was baffled.

That didn't necessarily contradict his thesis, though. "I'm of the opinion that it must be other things, like the movement in the music. Like the rhythmic aspects. Timbre is underrated," he said. "Particularly in pop music, there are a lot of changes that happen in the timbre, in the way that people produce music and then in the mixing and production of music that add lots of emotional colour. And I'm sure that's all there for you."

"Come Undone"

"Monkey Man" played a crucial role in my last year at McGill. The old Rolling Stones song was our "psyche tune," the track my roommate and I played—cranked—just before we left our apartment in the student ghetto and headed out to a party or a bar. We finished our beers, did a final bong hit, and screamed (and jumped around) along with Mick Jagger and the boys. That was our way of getting psyched up for the night ahead. As soon as the song ended, we flicked off the stereo, bounced down the stairs from our second-floor apartment, and bolted outside to the streets of Montreal. To this day, that song still pumps me up.

For most people, music is about emotion. Sometimes we want peaceful music; sometimes we want to dance. According to my friends, I like a lot of depressing music. Some of them call it "slit your wrist music" or "music for ending it all"—they consider Lucinda Williams a major offender—and they often give me grief about it. I also love to listen to upbeat music, angry music, meditative music, mournful music. I have emotions, and music helps me emphasize them. Or mitigate them. If I'm angry, I might want to listen to some angry punk so I can really indulge my fury, or I might want something cheerful to calm me down. If I'm bummed out, I can wallow in my sadness with some "slit your wrist music," or I might want to get happy. The beauty of music is that it has the power to do all that. And that's why I love it.

My diagnosis made me wonder how any of that could be possible if I was unable to hear music properly. Either there was something weird about me, beyond the amusia, or there was something wrong with the way we think we hear music. I couldn't doubt my diagnosis, but I could doubt the emphasis our society puts on pitch.

Here's the science: the basic unit of music is the note and we use pitch to indicate how high or low a note's frequency is. We measure frequency in hertz, or cycles per second, and middle C is 256 Hz.

Here's the culture: in the West, we've decided that accurate pitch is a fundamental marker of good singing and, along with rhythm, the crucial perceptual attribute of good listening. "I wouldn't even say pitch and rhythm are equal partners in Western music," said Frank Russo. "Pitch rules." That's probably because most of the rhythms in our music are fairly simple, which means we devote even more focus to pitch. At the expense of everything else.

I once asked Isabelle Peretz: "Do you think maybe you are overstating the importance of pitch in the way we listen to music?"

She seemed a bit surprised by the question. After expressing her doubts that my suggestion could be true, she said, "No, I am not overstating it. It's not only me, it's the whole field."

But some researchers in the field, and not just Russo, are increasingly open to other possibilities. "There are so many things that go into music other than fine-grain pitch discrimination," said Psyche Loui. "That ends up being the most studied aspect, but it doesn't mean it has to be."

Sometimes—in choral singing, for example—it really would be bad if the singers were out of tune. But in a lot of popular genres, singers have a bit more freedom from the pitch police. The obvious, most cited example of a tone rebel is Bob Dylan, who made his first album in 1962, enjoyed revered status in the '60s, maintained enormous popularity in the '70s, and still records and performs today. His success has not been because of his ability to nail all his pitches. But his voice has lots of character. In fact, perhaps that's part of Dylan's appeal: listeners may think that if a singer sounds like that, he must have something important to say. As *New Yorker* editor David Remnick puts it, "Dylan has something better than a 'good voice.' He has a true voice. He has a voice that brings out what truth there is in a song—particularly his own."

To be honest, I can't really tell that Dylan isn't hitting all of his pitches. That doesn't mean I think he's a virtuosic singer—I don't—but I can't pick apart one of his performances noting where he's, say, a bit flat. So that's an unexpected bonus for me. But I also have trouble discerning the difference between minor and major keys. Songs in major tend to sound brighter and happier, while those in minor sound darker and sadder. If I listen to examples of each, one right after the other, I can hear a difference, but it's subtle. If I go onto YouTube and listen to a version of R.E.M.'s "Losing My Religion" that has been digitally reworked to change it from A minor to A major and then listen to the original, the difference seems so small that I'm not entirely confident that I could tell which was which in a blind test. Worse, I always thought of that song as poppy and upbeat, despite the melancholy lyrics. Maybe that's why I like "music for ending it all" so much.

But the minor is sad, major is happy idea is a function of culture. Many of our theories about Western music developed in the 1600s, making them 400-year-old ideas about something we've been doing for more than 40,000 years. And we still cling to those ideas even though music has changed so much in the last four centuries, and especially in the last six and a half decades, since the birth of rock 'n' roll. Although I am tone deaf, that doesn't mean all notes sound the same to me. And the falsetto that Bon Iver's Justin Vernon uses in a lot of his songs, for example, sure doesn't sound anything like Barry White's bass-baritone. It's the small differences I can't hear; that's why the scientists refer to amusia as a fine-pitch discrimination disorder. Still, describing exactly what I hear when I listen to music isn't easy because I don't know what I'm supposed to hear or how other people hear it.

Long before I started my lessons with Micah Barnes, I met Gillian Turnbull. As an ethnomusicologist, she studies music and culture, including the way we use and experience music. At the time, she was finishing her Ph.D. thesis on the roots music scene in her hometown of Calgary — making her a doctor of country music — and teaching at Ryerson University. After a friend who'd taken one of her courses introduced us, I arranged to meet her for lunch at a Toronto café to talk about my interest in singing. I hadn't been tested yet, but I already figured I might be tone deaf and I wanted know if she thought trying to learn how to sing was a crazy idea.

"What first attracts you to a song?" she asked me as we munched on grilled panini.

The question took me aback because I hadn't thought about that before. I mumbled something about the lyrics, which I immediately realized wasn't true. I quickly corrected myself, pointing out that I usually need to hear a song a few times before the lyrics make an impression. Then I tried again: "Sometimes I think I hear music through my gut, rather than my ear."

While that statement clearly suggested a fundamental misunderstanding of human physiology, that didn't mean there wasn't some truth to it. And not just for me. When I was a kid, there was a popular television show called *American Bandstand*. I didn't really watch it much — and was surprised to learn that it ran from 1952 to 1989 — but we all knew about its Rate-a-Record segment, in which host Dick Clark played a new song and asked teenagers to score it between 35 and 98. Often the resulting analysis didn't go beyond some variation of "it's got a good beat and you can dance to it" — so often, in fact, that Clark later said he rarely went anywhere without someone repeating that line to him.

And let's face it: most people don't spend a lot of time thinking about what they like or don't like about music. We hear a song and we dig the sound or we don't. That doesn't mean that people with musical training won't give more thought to what appeals to them and be able to offer more analysis about the structure or the playing of the various instruments. And even people who don't play guitar may appreciate a face-melting guitar solo.

When he was testing me at BRAMS, Sean Hutchins, who has a lot of musical training, told me how he listens to music. "What I like are interesting harmonies, those jump out most at me," he said, adding that he also likes interesting chord progressions. "That's not to say that I don't appreciate the nice turn of phrase, but I think harmony's probably the most important aspect in my musical listening." For most of us, though, our initial judgement is usually an instinctive one — a gut reaction. It certainly is for me.

Our response to a tune often changes with more listens, of course. We pay more attention to the rhythm, the arrangement, the instrumentation, the harmonies, and yes, the lyrics. Back when we bought and listened to music by the album, my favourite song on a new record would often change the more I played the disc. After a few weeks of heavy rotation, a track I'd barely noticed at first would suddenly be the one I wanted to hear the most. And I think all serious music fans have, at one point or another, described an album as needing several listens. The first time I heard Wilco's *Yankee Hotel Foxtrot*, for example, I wasn't sure what to make of all the studio experimentation on it. But today it joins The Clash's *London Calling* as one of my two favourite albums of all time. Part of this is simply exposure. In *Guitar Zero*, Gary Marcus notes that just listening to a song tends to make us like it more, which he sees as an example of a phenomenon called "liking for familiarity."

Although I feared that no amount of familiarity would make anyone like my singing, Turnbull was supportive of my plans. After my first meeting with her, she offered to help me with some ear training. She was living in a basement apartment in Toronto's Annex neighbourhood back then. When I arrived, she had her keyboard set up in the living room. She played a few notes from well-known songs and I immediately recognized them. We also did breathing and relaxation exercises before doing some pitch matching. Afterwards, she gave me four songs to practise singing, and sent me some MP3 files of pitch intervals. I was to work at trying to match them. I did practise a little, but lost heart when Carmen said there wasn't much difference between my *do re* and my *do fa*. Next, I bought an electronic tuner like some guitarists use and tried to sing individual notes using the readings on the tuner as feedback.

Through no fault of Turnbull's, the couple of sessions I had with her didn't lead to anything. Still, we became friends and, over the years, we've had many conversations about music and singing. She was convinced I wasn't tone deaf before my diagnosis and, afterwards, she was skeptical that I was tone deaf in any way that really mattered. When she worked with me, I was able to hear the movement between pitches and to approximate that movement when I tried to replicate it. Although I didn't hit the exact pitch that she played, I was within the range of that pitch. "There's something being lost in the transmission of hearing and reproducing," she told me, "but I think you are hearing it."

Her ethnomusicologist's take is that so much of it is cultural. "To me, it's totally strange that tone deafness is even a thing, because there is so little emphasis placed on pitch in any popular music today." Psychologists think that you either can or can't hear something based on the physical make-up of your brain, she said. Ethnomusicologists ask, What have you been exposed to? What have you heard all your life?

One day early in 2012, Turnbull had an idea for an experiment. "Write down your fifty favourite songs," she told me. "Don't spend a lot of time on it, just write down the first fifty songs you love that come to mind." By asking for my gut reaction, rather than a carefully crafted list, she hoped to get a sense of what I responded to emotionally. Overthinking my picks could mean leaving out embarrassing guilty pleasures or including songs that might seem impressive in an attempt to cultivate a certain musical identity.

This was a surprisingly hard assignment. And even if I cheated a bit by checking the most played songs on my iTunes for some ideas, as soon as I hit send, I had my regrets.

Since it was only fifty songs, it was only a slice of my collection, so some genres I like weren't even represented. And why had I picked "Lay Lady Lay" instead of "Like a Rolling Stone" or "Just Like a Woman" or even something from *Blood on the Tracks*? Was it because "Lay Lady Lay" was the first Dylan song I remember hearing, and loving, on the radio as a kid? If that was why, then Turnbull was fine with it. That was the kind of emotional connection she was looking for.

Later, I started to think of the songs and artists I'd inexplicably forgotten. How did I miss the Talking Heads, one of my all-time favourite bands? For sure, "Take Me to the River," the brilliant cover of the Al Green song that's on *More Songs about Buildings and Food*, should have been on my list. Not just because of its brilliance, but because it evokes such vivid and good memories of a certain time in Montreal. A couple of later songs from the Talking Heads — "Life during Wartime," "Once in a Lifetime," or "Burning Down the House" — could easily have been on my list as well. Unfathomably, I'd also missed David Bowie. And Nirvana. And many others.

Then I had more second thoughts: I'd picked "Stuck between Stations" even though I'm pretty sure my favourite Hold Steady song is "How a Resurrection Really Feels." And I'd chosen Garland Jeffreys's "Coney Island Winter" from the new-at-the-time *King of In Between*, when I should have picked something from *Ghost Writer*. That was probably a case of recency bias, the tendency to pick more recent examples, especially when creating lists such as All-Time Greatest Hockey Players or Top 50 songs. Indeed, my list was heavy on music from the late 1970s and early '80s, when I was a student, and the decade or so before I drew up the list. In fact, I'd picked only eight songs from 1984 through 1997, but twenty-five songs from 1998 through 2011 (an excellent era for music, but still). Whatever the reason, there were so many good songs—Top 50 songs—I'd left off my list.

That wasn't a problem for Turnbull, who had what she wanted. Growing up in Calgary, her first concert was a New Kids on the Block show and she loves Rush, but our tastes meet at alt-country and country—she was, until recently, the Toronto correspondent for *No Depression*, the magazine devoted to roots, alternative country, and Americana. So she had some of the songs on my list and found the rest on YouTube.

I was skeptical about this idea, but she took my selections and evaluated them according to several broad categories that we can use to analyze music across all cultures, including melody range, melody movement, texture, vocal timbre, instrumentation, tempo, meter, genre, dynamics, and lyrical theme.

One song—"B-Movie" by Gil Scott-Heron—was such an outlier that she didn't even bother analyzing it. But she learned a lot from the other forty-nine songs. The feature that stood out the most was timbre. Most of the songs I picked had warm, generally mellow, almost whispery colours and textures that are dense and enveloping—sounds that aren't harsh or all that distinct, but mixed together well to create an overall warm feeling. That timbre was particularly prevalent in the voices I chose.

Such consistency might have been expected if all my songs were from one genre — if, for example, I'd chosen fifty country songs. By my selections came from a span of more than four decades, from 1969 to 2011, and included soul and R&B, a little bit of blues, a fair amount of alt-country and a little traditional country, and lots of rock, especially what some people call indie rock. So a broad enough range that people would say the songs didn't sound similar to each other at all. I mean, to me, "I Wanna Be Sedated" by the Ramones and Costello's cover of "A Good Year for the Roses" don't seem to have much in common. But Turnbull heard similarities, starting with timbre.

Some of her other findings: I like textural changes or builds in intensity (or both) within the first sixty to ninety seconds; medium tempo (many of my picks were in the 100 to 120 beats per minute range); duple meter, a time signature marked by beats in groups of two, usually 4/4 but sometimes 2/4 (though she pointed out that most pop songs are); and a recurring, syncopated riff in the guitar or bass. I also seemed to like songs with limited melodic movement in the vocal parts (that is, no big leaps or large ranges); vocal timbres that vary between slightly clear and slightly raspy, with a bit of extra growl or strain at times; and most of all, timbres and melodies that are speech-like (that is, not much of a difference between saying the words rhythmically and singing them). As for the lyrics, a lot of the songs were about someone longing for something such as a lost or unrequited love — a common theme in popular music — but surprisingly not many about the pain and torture of heartbreak. So more yearning and less hurtin'.

She also took a stab at what means the most to me in the music I love. In order of importance, she came up with this list: timbre (instrumental first, vocal second), texture (including build within the first minute and a half), subject matter, melodic range and movement, and rhythmic interest. That she had ranked timbre first seemed significant. Before I'd started my singing lessons and learning about music, I'd had only a vague sense of what it meant. Now, it was becoming increasingly hard to ignore.

In 2014, I was helping to make a show about tone deafness for CBC Radio's *Ideas* and I asked Turnbull to bring three of my picks to the studio so we could listen to them together. First up: "Alison" from *My Aim Is True*, and one of the three Elvis Costello songs on my list. Turnbull noted his expressive voice, adding that the little bits of raspiness in an otherwise warm timbre make it even more expressive. The guitar has a similar mellow timbre to it. This song has the syncopation common in my choices: there's a regularly occurring beat, but the rhythm of the other instruments and articulation of the melody happens off the beat, which creates a bit of a jarring effect. As we listened to the chorus, she pointed out the build in texture and noted that many of my selections start in a mellow way and then build to the chorus; some come back down again and some sustain that build right to the end. To her, the standout moment in the song is his singing of the name Alison. Costello puts all of his emotional expression in that one word, creating a great anchoring moment.

Next we listened to "I've Had It" from Aimee Mann's *Whatever* album. Turnbull pointed out the gentle guitar with the warm timbre that matches her voice, keeping all the sounds similar to each other, and she noted the syncopation in the guitar riff. In this song, Mann uses irregular phrasing: instead of singing in symmetrical melodies, she sings a short phrase and then a longer phrase and it doesn't quite match up with the symmetry of a typical duple-meter song. That's something I like, too, Turnbull told me. This song has only a gentle build in it.

While the build is something I like, it's also something most listeners like. In fact, it's probably one of the reasons we like music. "You want something to create excitement over time," Turnbull said. "If something is exciting right from the start and is this kind of onslaught of intense emotion, then you're maybe not always ready for that."

Then she asked me about the words. I admitted that I wasn't really sure what the song was about. (The explanation I attempted was, I realized when I looked up the lyrics afterwards, all wrong.) Turnbull noted that this was another song with a sense of wanting something to happen without being sure what it is.

As for the vocals, Turnbull noted Mann has a pretty but natural voice, and it's the voice of a normal person so it's something that we could imagine singing ourselves. "I think we attach a lot of expectation and hope when we hear a voice," she said. "So we think, okay, if I met Aimee Mann, could I talk to her? Yeah, there's something about her voice that makes her accessible, but also makes her a little bit mysterious."

For the third song, Turnbull reached back to the early '70s for "Satellite of Love" from Lou Reed's *Transformer* album. She said this song has conventional or symmetrical phrasing, but she didn't think anything else about it was particularly orthodox. For example, the background vocals really contrast with Reed's voice. And the music is not typical of most pop. There isn't much build, but lots of syncopation, especially on the piano. She drew my attention to "a brilliant moment" when the chords descend to a point where you'd expect some resolution. But they don't resolve, instead leaving you hanging until he resolves that tension by starting into the next verse on the correct note. "This idea of making you long for something and then finally giving it to you is the prevailing thing in this song."

Each of these tracks featured completely different singers, but Turnbull identified similarities in the way they deliver a melody. They tend to use somewhat unconventional phrasing, maybe cutting things off faster than a listener might expect, and they may be slightly off time or not symmetrical. And they all have a recognizable timbre that's warm in some way, though that warmth is different with each one.

But there was more. "Okay, well, I'm going to say something that you maybe won't like very much," Turnbull said. "But you have a lot of people on your list who I think generally the pop music audience would say can't sing." These included Lou Reed, Bob Dylan, Neil Young, The Clash, the Ramones, The Pogues, and even Patti Smith, who is idolized for many things besides her voice. (As Barnes said to me about Smith one day when our conversation turned to bad singers, "I have every album she's ever done—it's not because of her singing. It's in spite of it.") And Turnbull, who had analyzed her own Top 50 songs, admitted that many of the men on her list were also not great vocalists.

The only virtuosic singers I had picked were Neko Case and Marvin Gaye. In between, though, were several good but not great singers, including Gram Parsons, Jeff Tweedy of Wilco, and Costello. "So you hear these wavers in their voice, hiccups, cracks, things like that," Turnbull said. By delivering natural, rather than perfectly polished, performances, these vocalists possess a certain vulnerability and feel authentic. "They're letting their emotional selves be exposed through that singing."

Although most of my picks were by artists who were well known or at least well respected within their different genres, none fall into the mainstream pop category. No Beyoncé, no Shakira, no Lady Gaga. These singers, even though they have distinct voices, do sound like each other. "There is this sound that's fostered by the big machine of the music industry. What they expect will sell," Turnbull said. Costello, Mann, Reed, and the other artists on my list were not part of that. They have authentic voices that the industry views as not commercial and not easy to sell. "So that says a lot about your taste. You like things that maybe are happening just under the radar a little bit."

That my favourite singers seemed genuine and had speech-like delivery didn't seem like a coincidence to Turnbull. "My impression is you listen to something like this and you think you could sing it. And I wonder if that's the case with many of your first choices, that you would put it on in the car and you would sing and you feel like you did a great job because you believe that you are not a good singer."

This was a charming idea, and it might be true on some unconscious level—but only on an unconscious level. I don't hear well enough to know if a song is hard to sing or not. And anyway, when I am in the car alone, I don't worry if I'm a good singer or not. I just sing.

Though I'd asked her to bring three songs, Turnbull had selected four. The last was one she thought represented all of the things I liked. It was "Come Undone" by Isobel Campbell and Mark Lanegan, from their 2010 album *Hawk*. Turnbull pointed out that I have a weakness for duets, especially when the two singers have different timbres that work well together. That's certainly true with this unlikely pair. Campbell, formerly of Belle and Sebastian, sings in an angelic but ethereal whisper. Lanegan, formerly of the Screaming Trees, delivers speech-like singing in a deep rasp that suggests a life full of too much of the wrong kind of fun. "There's almost nothing to her voice; it's like something that appears for a second and flits away," Turnbull said. "And his voice is very deep and so we end up with this whispery quality as a result. They're quite far apart in range but something about that combination makes it a warm timbre overall."

Since Turnbull teaches courses in popular, world, and classical music that attract both musicians and fans, I wondered how the way I hear and respond to music compared to the way her students do. She suggested that any differences could probably be better explained by generational forces than by brain deficits. Electronic music has been such a dominant part of the soundtrack of her students' lives that they tend to focus on beat: how fast a song is going, how prevalent the beat is, and what instrument it's played on. They focus on lyrics as well. The first few times she played The Kingston Trio's version of the traditional murder ballad "Tom Dooley" in the folk segment of her pop music course, she was confused when her students gasped and laughed. Eventually, Turnbull realized that a woman gets stabbed with a knife in the song's first verse. She had never noticed those lyrics before.

And timbre is huge for them. They don't appreciate the early-twentieth-century music she plays for them, because they don't the like the quality of the recordings. That's another kind of timbre and one so different from the cleanly produced music — often overproduced, with a heavy reliance on Auto-Tune — that they've grown up with. Timbre is also one of the concepts that they grasp quickly. Turnbull will play something and ask her students to describe the timbre of the singer and the timbre of the instruments, and they are able to give her good, solid answers. But they struggle with similar questions about meter, harmony, and pitch (though by the end of the semester, they do start to hear those things). "It takes a long time for them to figure that stuff out, but timbre is an immediate reaction. You know the quality of a sound when you first hear it."

After analyzing my Top 50 songs, Turnbull was convinced I was hearing a lot more than pitch (or, perhaps more accurately, hearing a lot instead of small changes in pitch). "There is something about the overall sound of a recording and somebody's voice mixed into that that is attractive to you, and there is consistency across all of those songs in that way," she said. "You do like the warm timbres, the big orchestration, the build and texture. All of those things are attractive qualities to you that override any particular melody or the way that melody is delivered in terms of being exactly on the pitch."

Turnbull believes all listeners, including me, are attracted to a lot of different qualities in music that we don't necessarily think about. "So what you're doing as a listener is what's natural in all of us. You're responding to the whole package of the music and picking out a variety of things that are attractive to you — and mainly that is timbre — as a way to connect to the song."

Side Four

Unlocking a Surprising Secret of Music

"Teacher Teacher"

When my friend Andy Malcolm, a musician, songwrit-er, and schoolteacher, was dying of cancer, I took him to an Elvis Costello concert. Like me, Andy was a big fan so it was the least I could do. During pre-show drinks, I revealed my secret scheme to learn to sing and then perform at an open mic night. That wasn't ambitious enough, he insisted. He wanted me to write a song for Carmen and then sing it. He offered to help me with the songwriting and accompany me on guitar. I liked that idea, at least as a dream scenario.

But after my diagnosis, which I received shortly after Andy died, I knew that was just another dream that would never come true. I needed to be realistic: It was unlikely I'd even be able to pull off karaoke without getting booed off the stage.

I took another hiatus from my lessons. More than a year this time. They were expensive, hard work, and frustrating, so I had lots of excuses. But eventually I decided I needed to go back to them and make one final push to finish my quixotic quest to sing in front of other people, no matter how it ended.

What Micah Barnes wants in his students is a willingness to be vulnerable, to open themselves up to a difficult process, and to work hard. Since his clients are mostly pros and singers who want to be pros, and he's expensive, they take it seriously. As he said, "No one is going to fuck around at a hundred dollars." Still, I needed more than an improvement in my work ethic. I needed an attitude adjustment. Early on, in particular, he had to deal with my resistance to singing, which included a reluctance to work between classes. Practising something you're terrible at is definitely no joy. "So, really," he said, "it was psychological warfare to try to get you into a place where you were open to this very challenging experience."

While most students aren't as difficult as I am, singing instruction is a big industry. A lot of people can't sing well but dream of being able to sing well, and there are lots of coaches willing to take their money. "People are desperate," Barnes said. "There's an ego thing around singing." Most people go to what he calls the "coach on the block." If I'd gone with one of those, I'd probably have given up long ago. Barnes and I had been making progress and, though it was slower than I would have liked, I wasn't planning to change the plan when vocal coach Elissa Bernstein emailed me after she heard the radio show on amusia I'd helped make: "I still haven't found anyone who couldn't learn so am intrigued by your story. I love a challenge. And I especially love proving that everyone can learn to sing (and act and dance). In fact, it is my life's mission to dispel the myths about talent and the performing arts, having been one of the 'have nots' for so long. If you're interested in trying another approach, I would be happy to take you on to try to prove your experts wrong." How could I not at least speak to her?

Bernstein's story is the classic one — or at least it starts out that way. As an eight-year-old Jewish girl in an ecumenical school, she was preparing with her classmates for the Christmas concert. The teacher, Sister Josie, said something sounded a little off and asked everyone to sing alone for her. When Bernstein sang, the nun had found her culprit. She told the girl to just move her lips. "I think most people love to sing, but I was one of those who really, really loved it," Bernstein said. For the next ten years, though, she wouldn't even move her lips. "I just believed that I couldn't sing so I wouldn't even try. I didn't even sing in the shower."

When she was nineteen and interested in acting, she started singing again. "I just sensed that I had a voice, and no matter how many people continued to tell me I didn't have the talent or I wasn't musical, I never let go of that." It took her years to find the right technique. That meant years of trying different teachers — she went through a dozen — most of whom made her feel like she should give up all hope. Even when she learned to sing and developed a good voice, she sensed that people considered her somebody who could never be a great singer. So her next goal was to break down the skills to really understand and master them. In 1997, she finally found the right coach: Steven Lecky, who had developed a technique called the Vox Method for actors, singers, and speakers.

Suddenly, it all seemed so simple to her. "What makes great art? Is it magic? Or is it great skill?" she said. "I've learnt and I've developed that great skill, and I believe that most people—and this goes against the establishment—most people can achieve a very high level in an art form."

Today, Bernstein is not just a singer, she's a vocal coach and a partner with Lecky in the business. Teaching in Montreal, where she lives, and around the world via Skype, she trains people of all ages, both professional and amateur, and in all styles, including opera, jazz, and pop. A lot of them have been told they can't sing.

So we arranged to Skype, and she gave me a short lesson as an assessment. Her approach is that singing is, fundamentally, just an expression of speaking. It's the same function of the voice. "Where singing becomes hard is because people have such a misconstrued idea about what singing is," she said. As an example, she told me about how she took a two-year-old through some of her exercises and the child was able to do them immediately. Other students—who are older and have a lot of tension and psychological baggage—can take up to two years to do them. "She didn't know enough to do things differently in those exercises than how she speaks," Bernstein said. "Everything is the same for her—the voice just functions naturally."

Bernstein believes people who think they're tone deaf are extreme examples of how we misinterpret what singing is. "They are so hell-bent and worried about hitting pitches that they're not following the natural function of the voice. And that is the spoken function. Can you just say 'wow'?"

"Wow."

"Now, I want you to think about that in terms of the trajectory it takes, so do it slowly."

"Wow." But I didn't do it slowly enough.

She said it again, drawing out the word, to demonstrate. And I copied her.

"That's a nice sound. Hey, you've got a good voice there," she said cheerfully. "All right. Can you sense that it starts somewhere and it goes somewhere else?"

"Yup."

"So that's how our voice works: Hey! What are you doing? What are you *doing*? I didn't say that. Wow, look at the day." She said these expressions in an animated voice. "So voice works like that: it comes from rest and it goes 'wow.'"

Unlike most coaches, including Barnes, Bernstein doesn't get too hung up on breathing. That's a small part of it, she acknowledged, and she does have to watch for shallow breathing. But she doesn't believe singing takes as much air when the technique is good. "Breathing is one of the myths of singing," she said. She worries more about tension (though Barnes would argue that poor breathing and tension are related). Since most people without training are much better speakers than singers, she aims to take advantage of how the voice is already working better than they think.

Building from my speaking voice made sense, I told her, but my problem is I don't hear pitch well. "How do I deal with that?"

But she was having none of it. "You obviously hear because you speak in a musical way, you have contour to your speaking line." She asked me to repeat that phrase — "How do I deal with that?" — several times after her. Then she said, "You just matched pitch perfectly. So you don't have a pitch problem," she assured me, "you have a co-ordination problem."

Inevitably, she asked me to sing. I started into a song Barnes and I had often worked on: "Amazing grace / How sweet the sound..."

"Okay, good, all right. Now, first of all..." she started.

"Thank you for interrupting me," I said.

"Now, I want you to say 'Amazing grace.'"

"Amazing grace."

"How sweet the sound."

"How sweet the sound."

"Exactly. You just matched my pitch," she said. "When you speak it, you do it perfectly well. Do it again. Amazing grace."

"Amazing grace."

She said it again, this time in a sing-songy voice.

"Amazing grace," I sang.

"And you just sang it perfectly in tune." Then she spoke: "How sweet the sound."

"How sweet the sound."

Then she sang it.

I sang: "How sweet the sound."

"Okay, you switched back to your old singing. And that's where the work is: to get you off the" — here she sang badly — "'How sweet the sound' into" — here she used more of a speaking voice — "'How sweet the sound.' Speak it for me."

"How sweet the sound."

"That's it and it's the exact same thing. Listen to me: How sweet the sound," she said. "How sweet the sound," she sang, then added, "Speak it and sing it."

"How sweet the sound. How sweet the sound."

"Exactly. Exactly. Much closer. Now, again, do this: 'How...sweet...the...sound...'" She placed one hand on top of the other and as she said each word, she raised the top one up and then brought it back down.

I did as I was told.

"Exactly. And that's how pitch works."

We kept going until she asked me to draw a staircase on a piece of paper. Then I spoke the start of "Amazing Grace," pointing to different steps on my staircase drawing for each syllable: "A-may, a-may."

"There you go. Perfect. You got that interval. You are not tone deaf, okay."

Later in the lesson she asked me to repeat "One, two, three, four."

"So your pitch is perfect when you speak it. Why? Because you're using your natural voice, which knows how to find pitch if you use a natural function."

We did it some more.

"And that's perfect. You have a co-ordination problem. You're using the wrong approach to singing."

"You make it seem easy."

"It is easy — if you have the right approach."

"So what do we screw up when we move from speaking to singing?" I asked.

"We revert to this straight, shoving, pressing sound.

We're trying to hit pitches, we're trying to get a flow because we think singing is about flow," she said. "But the flow will come if you get a natural release onto the sound... I don't make my voice float. I develop it so it resonates on its own, the way it is meant to function. This is an instrument. You can't play your cello by hitting it against the wall and squeezing it and expecting a sound. You have to play the instrument the way it's designed to be played."

"But," I protested, "I do have problems hearing pitch."

"Tim, you just repeated this" — she played a note pattern on the keyboard — "perfectly, which is not easy. A lot of my students can't even do that."

She had me sing "One, two, three, four, five," while stepping forward one step for each number.

"It's hit and miss because you keep reverting to the singing. And that's normal. The brain will always come in and try to shake things up, so what you need to do is focus on these sounds."

I explained about the testing the researchers had put me through. I said I understood that sometimes I can hear on a subconscious level, and that's probably what was happening when I was able to match her voice. "I've been tested in a lot of labs," I insisted. "I am tone deaf."

"I couldn't care less about what the labs say," she said. "Your voice and your brain know how to match pitch. So I'd say, stay away from the doctors and get yourself a technique — whether with me or someone else — that works from the spoken sound that allowed you to match those pitches. If you can match those pitches, you can sing. Whether or not you know if it's higher or lower, who cares?"

Before the end of the lesson, we spoke-sang "Amazing Grace" again. "You have perfect pitch recognition. You drive me crazy," she said before singing, "You drive me crazy." She continued: "Now, you don't sing like that and that's why you're still not a good singer. That's why you don't sound like you can sing. But I know that you can."

Bernstein's background is law and science so her approach to life, she explained, is all about solving problems. And her big disappointment is that so many people don't want to hear that they can learn to sing. She recounted an anecdote about a renovator who'd heard her singing opera the day before and told her it had given him goosebumps. She told him she could teach him to do that. "No, that's impossible," he said. "That's a God-given gift." He didn't want to hear that it was as mundane as understanding the technique and then practising it. "People want to believe in magic. But the production of sound is not magic, it's all skill. And isn't that wonderful? Or do you want to believe in the tooth fairy and magic?"

I was intrigued by her approach and, later, when I listened to my recording of our session, I could hear how much my "Amazing Grace" had improved using her technique. But I was also improving with Barnes, and it didn't make sense to abandon him. Even Bernstein reluctantly conceded that switching singing coaches is like switching hairdressers (an analogy I grasped well, if only on a theoretical level).

Larra Debly (who performs as Larra Skye) is another musician who has set up a teaching business. She even hires others to work for her company, Sweet Music Lessons, which offers in-home lessons in Toronto. She hears a lot of different voices and a lot of different obstacles to singing. The biggest is often confidence. She can empathize, because while she's totally comfortable singing on stage, she's terrified of speaking in public. Even at her shows, she feels awkward talking between songs. Among her adult clients, some of the least talented have the most confidence, but the kids will play one wrong note and focus on that. Some of them are really hard on themselves and they're just ten years old. "Don't worry about mistakes," she advises them, "it's not the end of the world." Or she'll say, "Hey, I played a show on the weekend and I made a lot of mistakes and it was okay. Nobody could tell because I just smiled all the way through it."

As a proud member of the precariat, I have fashioned a career that involves a mix of jobs. One of them is as a journalism instructor at Ryerson University. Because I teach, I'm always interested in how others do it. Whenever I take a class or a workshop—or singing lessons—I'm always thinking about two things: what the teacher is teaching, and how the teacher is teaching it.

Inevitably, this was a frequent topic of conversation during my sessions with Barnes. Early on, I told him, "I was thinking after the last class that I always come out of here feeling so much more positive about this project, rather than feeling beat up. I mean, I know it's hard. It's the same thing I tell my students: 'This is hard, don't kid yourself. I've never told you this was easy, I've never suggested this is easy. I've said it's hard.' Just the way you teach is very encouraging and tolerant and forgiving, which is good because I have a long way to go. If you had really high expectations, it would not work, I don't think."

"I have really, really easy expectations," he said. "My job is to figure out how you learn."

At later lessons, he'd often talk about how working with me was good for his patience. He claimed he wasn't a naturally patient person, though he always was with me. (I did see him chew out a client who'd phoned in the middle of one of our sessions to cancel, not for the first time, at the last minute and beg not to be charged. He was not a man I wanted to cross, I realized.)

Although developing his patience might have been an unexpected benefit of working with me, it didn't really explain why Barnes wanted to take on such a bad singer. He does enjoy working with a musician who is recording an album or preparing to go on tour, or a television star who suddenly discovers her part calls for her to sing. "That's really fun," he said. "But I believe music should be in everybody's life. I'm not a snob about that."

As a music teacher, my friend Tyler Ellis definitely agrees with that. He's always sung, and he remembers a family road trip across the country when he was eight or nine. His mom kept a songbook in the glove compartment of the Toyota Crown Royal sedan, and they sang all the way. His mom even sang along, blushing only slightly, when his dad led the family in renditions of saucy engineering songs or the risqué, innuendo-filled ditties of George Formby. Ellis grew up to be a musician. He still regularly performs around town, with and without a band, and records albums. But his day job now is teaching music to kids aged four through twelve at Morse Street Public School in Leslieville, a rapidly gentrifying Toronto neighbourhood.

In the classroom, his attitude is everyone—including the shyest or least confident kids—join in and make music, even if it's just to bang a tambourine. Often that's enough for them to forget that it's such a nerve-racking experience. Usually, he can get even the most timid pupils to pick up the bass or another instrument. Ellis strongly believes that everyone can—and should—make music.

I visited him one day while he taught a group of students in grades five and six. He had decorated his large classroom with posters of musicians ranging from Bob Marley and Bob Dylan to Nelly Furtado and Taylor Swift to Dizzy Gillespie and k. d. lang. A copy of David Byrne's *How Music Works* and some figurines of musicians sat atop the upright piano. The students gathered on and around a large carpet in the centre of the room—some of the older ones grabbed chairs, the rest sat on the rug—facing a projection screen that displayed the lyrics. Even when everyone knows the words, he keeps them up there because it helps the singers stay focused. He played the keyboards for the first couple of songs and then one of the kids took over when they did "Count on Me," a Bruno Mars number.

From that point on, Ellis was free to walk around and use proximity—a teacher's trick, he called it—to encourage the reluctant singers. Almost all of the students were taking part. But some kids come "pre-loaded to sing their hearts out," and some were "not really using their instrument quite the way they could" or became distracted or for whatever reason weren't singing. So he stood near these kids and sang, which improved their focus and gave them a little prodding. "If you sing in their ear a little bit, they seem to like that," he said. "If I'm willing to do it, it can't be that weird."

Ellis would never tell a student to just mouth the words. His background is songwriting and campfire singing. He's not really interested in putting together a choir for competitions (though he knows excellent teachers who do and he's impressed with the results). In choirs, people have specific parts, such as alto or soprano, and a singer who can't hit the right notes can be a problem. If you put a soprano who's out of tune with an alto who's in tune, for example, you no longer have harmony. But Ellis's approach is unison singing. Everyone sings the same part, which is much more forgiving for those who struggle with the pitches. It also means people are more likely to jump in. It's more like a campfire singalong. "When we do a show, people are not looking at—or listening specifically to—one student who might be off-pitch," he said. "People are looking at the energy of the group, and the joy that they have producing the music."

Sometimes it can be hard to tell if students are reluctant because they're shy or because they believe they're bad singers and they're nervous about being judged. He's down on the culture of judgement we live in, one especially fostered by TV shows such as *American Idol* and *The Voice*, and is discouraged by the influence it has on his students. "If you're sitting beside your neighbour, and you open your mouth, you're putting yourself on the line," he said. "It's a big risk when you're singing and you worry about wrecking it for other people, or someone going home and saying,

'I heard so-and-so singing and they're just awful.' It's such a big risk."

I asked Ellis if he'd ever taught kids he figured really were tone deaf. He couldn't say for sure if they would have failed an amusia test, but he'd taught students for whom it was beyond his capacity to bring them up to singing on pitch and help them become really strong singers. In those cases, he concentrated on doing what he could in the limited time he had with them. He focused on songs that have a lot of repetition and simple do-mi-so melodies and tried to bring them into the group. And he spent a lot of time beside them when they were singing, just trying to get their pitch to meet the other voices.

"It's a continuous process of trying to make them aware, so that they can hear the sounds in their head the way that many just naturally do," he said. That's always led to some improvement. "Some kids are meant to be athletes and some are meant to be performing singers. It's the attitude they bring to it. If you bring everything you have, then it's going to be great."

Ellis is someone who truly loves singing and says he can transport himself anywhere with a song. "It's like skipping," he said. "If you're skipping, then you're probably happy. You don't see sad people skipping." Unfortunately, many kids don't see singing as cool, certainly not as cool as sports. Even though they all look up to famous pop stars, somehow they don't want to sing themselves—at least not with others.

But Ellis takes the long view and compares it to the hockey coaching he's done. He knows neither his daughter nor his son will make the National Hockey League, but they both still love playing. "That's what the game plan is. Take this thing that everyone already loves, don't wreck it for them, and let them go off and have it as part of their life," he said. "It's all about happiness, and singing is a big component of that. If you don't have it, you could still be happy, but it's such a great tool to have in your tool belt. If you can give people that, then job well done, really."

"If I Only Had a Brain"

When I returned to my lessons in the spring of 2014, after more than a year away, most of it spent feeling guilty, Barnes's studio—the living room of his apartment—had a new look. He now had a Weinbach baby grand piano, and he'd replaced some of his old furniture with more modern pieces including, beside the new piano, a white cushioned chair that let me sit up high, as if I were on a bar stool. I liked that.

From when we'd first started to work together, Barnes was always trying to get me to relax. Fear affects the body, he assured me, and my body tension was pushing me out of tune. That's why he was always hectoring me about my breathing and, at least in the early days, making me lie down on the couch or the floor or do weird movements. Before I started using the iPod Nano for practising, Carmen could hear the recordings of our sessions, which included both our singing and our conversations. She was looking forward to meeting Barnes so she could ask him how he managed to get me to do whatever he wanted.

Eventually, I grew more comfortable with Barnes and the techniques he was teaching me and the whole idea of learning to sing, but I was still tense about hitting my pitches. So he had to find different euphemisms to cajole me into singing in tune. "A little more courage on that last one," he said during one session, "i.e., pitch."

I was now working hard between sessions. But each lesson took so much concentration that Barnes could see that I would lose my focus by the end of the hour, and my ability to hit pitches would deteriorate. One day, after we'd been working on "Amazing Grace," he told me I was losing it, so we stopped. But he was still encouraged by my progress. "Places where you're stuck, you solve it," he said. Then he added, "Without a doubt. The practice is really changing it — really changing it. I find that you're tense about the whole thing, though."

"I guess because I'm not sure of the note, I'm tensing about whether I'm going to hit it. I might know it's supposed to go down, but I don't know how far down so I'm stabbing at it."

"Some of it is simply relaxing. I can see it."

"I think I have to let my subconscious take over."

"People who are not willing to swim in that ocean can't be good singers," he said. "A singer goes into a concert or recording studio and doesn't know, can't control it. That's a really huge guiding principle to our lives. It's why artists have really complicated relationships to control. We want control and yet we never actually get it."

He was determined to get me to relax and trust my ears. In some ways I was a unique student, but in others I wasn't. His job often involves stripping away the psychological hang-ups that get in the way of good singing. He's always listening for where the muscles are tight, where the person is fearful, where the body is stopping the sound. That means a lot of his work is about removing tension. "What I'm listening for is what I can remove and therefore liberate the voice."

He can, with only a little prompting, get philosophical — maybe even a little mystical — about singing. He believes it is about liberating the spirit. "We've understood it as a cry from the soul. We've understood it as prayer. We've worked it into religious ceremonies. We've worked it into courting ceremonies. We use singing when we're happy. We use singing when we're sad. We use singing to survive our lives. We use singing to imagine a better life," he said. "It's really a way of tapping into the larger and deeper aspects of our existence here on the planet, and I believe it's a saviour, in many ways."

But on a more mundane level, I had a challenge that was bigger than it should have been: trying to find the ideal song to sing. Something not too difficult, but something I liked enough that I could sing it with emotion. One day, even before Gillian Turnbull had suggested that I liked songs by singers who aren't considered good singers, it dawned on me that the song I really wanted to perform in public was Joe Strummer's "Silver and Gold." I sent Barnes a YouTube link to it and asked if he thought it would be a good choice for me. His response: "Perfect. He's already singing out of tune :)"

I laughed at this, but the song means a lot to me. It's the last track on *Streetcore*, the solo album the former Clash frontman was recording with his band, The Mescaleros, when he died of a heart attack in 2002. A retitled cover of "Before I Grow Too Old," in Strummer's hands it's a song with a country-inflected sound and anthemic lyrics about doing life up right, and at full blast, before it's too late. Given his death before the album was even finished, it serves as his own unintended eulogy, especially since after the song ends, we hear a pause and then he says, "Okay, that's a take." It's the song I want my friends to hear at my memorial. But before that happens, it seemed to say all the right things about my own desire to learn to sing before I'm too old.

My mom, Elsie Falconer, is an accidental artist. And not just because she didn't start until she was about seventy. After a long, successful career as a real estate agent and broker, she wanted something to keep her mind active. "I had to do something when I retired," she said matter-of-factly, though she still works part-time. But other than photography, which she liked because it was simple and mostly involved snapshots of her family, she had never done anything artistic in her life. So she tried pottery. It was tactile, which appealed to her. After a couple of years, she realized that knowing about painting and mixing colours would help her pottery, so she signed up for an acrylics class.

Before long, she was hooked. And the bonus was that

while she could never figure out where to put her all her pottery, her paintings were easier to store. Though the truth is, the walls of her home are filled with her art. Today, she spends about nine hours a week at it, regularly taking courses and sometimes going away on artist retreats. She enjoys meeting new people, especially younger ones, but she believes she looks at art in a different way than her classmates. "I always want to figure out what the painter was trying to do, and most other people look at what the painting gives to them." And she looks at the whole picture rather than picking out details such as the brushstrokes— an eerie analogy for how many amusics, including my mom, hear music as a blended whole and have trouble picking out distinct sounds.

My mom was sitting at the kitchen table by the front window of her townhouse. The sun streamed in, filling her workspace with lots of natural light, as she sketched with charcoal on a canvas, prepping it for what would eventually be an oil painting of a Venice canal. She was working from a colour-saturated photograph my sister had shot on a trip to Italy. The detailed scene, with shimmering reflections of buildings on the water in the foreground and a long perspective in the background, didn't look like it would be an easy one to paint. It was going to take some time, but that was okay.

Not surprisingly, how my mom views paintings influences her own work. "My art is much simpler," she said as she sketched the outline of a docked gondola. "The other people are very detailed." I looked at some of her paintings in the nook behind me and could immediately see what she meant. Rather than discernible brushstrokes and detailed depictions, everything in her paintings has a softness to it. I asked her if she wished that she had a different style. "Well, it's all I could do so...and I like the paintings I do, so...that's what I have to deal with," she said, pretty much summing up her attitude to life. And other people like her art, too. Despite that, she doesn't think she has natural ability. Most of the students in her classes have at least taken an art history course in university, which she never did. But she likes doing portraits, and with hard work, she has really improved on perspective paintings.

My mother's ability to learn a new skill later in life — especially a skill she thought she had no real aptitude for — is a classic example of brain plasticity. And as Gary Marcus showed in *Guitar Zero*, adults can learn to make music (the book's subtitle is *The New Musician and the Science of Learning*). Meanwhile, in 2006 and 2007, Steven Mithen spent a year learning to sing. Unfortunately for the archeology professor who wrote *The Singing Neanderthals*, his singing didn't improve as much as he had hoped it would, but before-and-after fMRI brain scans showed significant changes in brain activity when he sang.

My mom's experience as a late-in-life artist is just one more reason why she's an incomparable role model for her son. But I already had my own example to follow. In the weeks before I turned forty, I started to freak out about getting old. I'm not given to panic attacks, and this was as close as I've come. I was halfway through my life and I had never done anything artistic (unless you counted my writing, which I didn't because it was my job. Besides, I'm just a journalist). So I decided to take up photography and, I think, became fairly decent at it. I've even sold some photos to publications. So why not singing, too? Unfortunately, my challenge goes beyond adult learning—I must find a way to route around the damaged architecture of my brain.

My maternal grandmother had a lovely voice, though my grandfather didn't. Neither did my mom, so she had never been encouraged to sing. Still, she considered taking up singing as a retirement activity even though she thought—correctly, as it turned out—that she was tone deaf.

That both my mother and I are amusic is not a coincidence. Isabelle Peretz believes amusia is a result of a genetic mutation and thinks it may be a mutation of more than one gene. Statistically, there's a 40 percent chance that an amusic's child will also suffer from the deficit. ("So I can blame it on my parents?" I asked Peretz. She responded, "Yes. Or your grandparents.")

She'd already spent several years on a large genetic study of amusia in families when she asked me to recruit mine to be part of it. So in the fall of 2011, I sent out an email with "Help a brother (and son) out" as the subject line. First, my mom and my sisters did the online test. Their scores were all in the low 70s, suggesting I wasn't alone. Then they gave saliva samples and had to sing "Happy Birthday" over the phone—first with the words, then using *la* for each syllable. They all lost the melody after the second *la*. The family dinner after this "singing" was full of cringing accounts of disastrous performances. (My niece and four nephews took part in the study, and two of my nephews also appear to be tone deaf.) That all five of my mom's kids are amusic is unusual. We are all convinced that my late father was tone deaf. Perhaps that explains why my family crushed the 40 percent benchmark. "All of them are amusic, which is very surprising," Peretz said. "I've never seen a family like that."

Another surprise for her is that amusics may perceive a lot when they listen to music. They're just not aware of what they're hearing, because there's a disconnection between unconscious processing and conscious awareness. That's a mystery she'd really like to understand. She also wants to know more about beat deafness, a less studied form of amusia. She identified and studied the first case — another journalist, as it turns out — who really can't dance to music. "He's unable to feel the beat," said Peretz, who has since found several more people with the same problem.

Most of all, though, she wants to find the genetic source of the disorder. By the spring of 2015, she had twenty-two families taking part in the genetic study. Convincing every-one in each family to take part, do the tests, and provide the DNA sample was taking time, which was endlessly frustrat-ing for her. "I suffer," she said about the wait. But, realiz-ing that she won't get everyone's participation, she doesn't plan to delay too much longer. The one good thing about the study taking so long is that in the meantime, other research into music and genetics has helped narrow down her search.

If she can identify the source of amusia, Peretz is hope-ful that it will help us better understand music and the human communication system in general. "If you find the genes, you find, really, the origins of music," she said. "So it's very exciting if we can find it." Maybe she'll even be able to settle the debate about whether music is evolutionary or just auditory cheesecake.

For all her excitement about the genetic study, Peretz doesn't believe it will lead to a cure for amusia. In fact, she believes we already know how to deal with it, though it's probably too late for me—just one more reason I regret not trying to learn to sing when I was much younger. "The cure will come from neuroplasticity," she said. If we identify children with the deficit early on, before the age of six, and immerse them in music lessons, she's convinced it will help in the same way that identifying dyslexics and giving them extra help improves their reading. "We have to try now. There is a big trend in the literature showing that people who do learn music early on have a cognitive advantage. If this is true, then there is a handicap for those who don't have it."

"Smells Like Teen Spirit"

By now, I had a good sense of what I could and couldn't do as an amusic singer. With hard work and a lot of feedback from Barnes, I could learn the contour of a song. But I was hit and miss on the pitches — and I couldn't hear myself well enough to know when I'd hit one and when I'd missed. So Barnes suggested a shift in strategy. I had to change my measure of success from the mathematics of nailing the notes — "I think we lose in that game," he said — to the new measure of my emotional commitment to what I was singing. If I'm emotionally committed, but singing all the wrong tones, I asked, isn't that still terrible singing? He didn't think so. "It may feel better to you, and it may feel like you can do it with less stress," he said. "That's a win, because then, eventually, as your ears improve, singing becomes something in your life, and that was my goal."

He argued that having an emotional understanding of the material gives a singer confidence. More than that, though, if I could communicate emotionally, the notes wouldn't be as important. "The idea that you have to be singing with pretty tones went out the window many, many decades ago, and singing is communicating," he said, citing Nick Cave, Lou Reed, and others as really talented communicators who don't have what anyone would consider pretty voices. "On the other hand, there are a bunch of fabulous voices that come parading out like lemmings, that don't communicate anything all. They just go, 'I can hit these notes really big and really loud and really pretty,' and it doesn't really mean anything. Singing is about meaning. What are you carrying to the humans about the human experience? Sometimes it's a haggard, worn-out, lost-in-the-woods cry that actually communicates more than the beautiful tones of a classically trained opera singer."

Barnes also believes in the healing power of singing. That's not something he says glibly or even lightly. When his partner René Highway, the Cree dancer and actor, died in 1990 of AIDS-related causes, Barnes relied heavily on music to heal or at least survive. The songs of Emmylou Harris, he once told me, were instrumental in getting him through his mourning.

For those who do survive, he thinks we can really hear who people are when they sing. That's not necessarily true when they speak, because they can choose their words and obscure their feelings. "But when you're singing, it's the truth," he said. "A lot of times as a coach, I feel called, in a powerful way, to work with people so that they have a deeper access to their true selves, and that's what motivates me. It's what I love about it, and there will never be an end to my exploring how to get to the core of a person, and allow them access on their own, to be able to communicate from that core, and communicate to the rest of the human beings who they are and what they care about. The best singing happens when we're not thinking, when we're just experiencing and translating our experience through song. There's such an uncomplicated kind of joy about that."

Researchers are finding fascinating and powerful links between music and health. They're using music in cancer treatment, pain management, and end-of-life care, bringing back memories to Alzheimer's patients, repairing the brains of stroke patients, and helping people with Parkinson's disease walk more steadily and quickly. As for singing, specifically, the health benefits include elevated mood, greater lung capacity, and a strengthened immune system due to lower stress levels. When we sing, our bodies react the way they do when we play sports, increasing blood pressure and producing endorphins. The Royal Conservatory of Music and Russo's SMART Lab have created a choir for people with Parkinson's that improves vocal quality, facial expressiveness, and the perception of emotion. No auditions necessary. The singers perform both happy, upbeat songs and sad, slower songs, while mirroring the choir director's expressions for each of those emotions. Such projects, and the increasing popularity and success of music therapy, suggest that perhaps Cervantes was onto something when he wrote "He who sings scares away his woes."

That's the attitude of one of Larra Debly's clients. Most

of her students are kids, but she also has some adult ones, and they all have interesting reasons for taking lessons. One man, who really loves music, has a lot of health problems. At their first session, Debly started showing him some vocal warmups. But that was more advanced than what he was looking for. "Listen, I'm not going to be Frank Sinatra," he said. "I just want to sing songs." His singing is improving, but that's not his goal. He just wants to have fun and get his mind off his pain.

The popularity of no-audition choirs suggests that a lot of people just want to sing. England's London City Voices, for example, is a community choir with branches in three different parts of the city and a repertoire that ranges from Queen's "Bohemian Rhapsody" to the South African national anthem to "If You Don't Know Me by Now" by Harold Melvin & The Blue Notes. In Edmonton, ·voice coach Eva Bostrand started A Joyful Noise in 2004. She added a second choir in 2008 and a third one in 2014; while each is suited to singers at a different experience and commitment level, none requires an audition. Most large U.S. cities have several no-audition choirs to choose from.

In 1996, Louise Rose, a former backup singer for Ray Charles, and three friends placed an ad in Victoria, B.C.'s *Times-Colonist*. They wanted to create a gospel choir for a Good Friday program and hoped to attract thirty or forty people. Enticed by the promise of no auditions, 326 people showed up to the first information session. Today, the Victoria Good News Choir—which still welcomes all aspiring singers, regardless of ability, without making them try out—has over two hundred members, though one woman never sings. The vibrations of the choir make her feel better. She gets to laugh with the others during rehearsals, and no one has a problem with that at all. Many members report that just being around other singers improves their mood. Plenty had been told they were tone deaf when they were young and believed they'd never be able to sing a note. Now they do.

My friend Ruby Andrew was in the Victoria Good News Choir before she moved to New Zealand. "What I found was that singing among 200+ people is an extremely powerful experience in itself—but also, that Louise was right, we could all sing," she remembered in an email. "I still think about our Tuesday night practices, which were held inside a church because that was one of the few places large enough to hold everybody. Singing with so many people literally sets up some magical physiological reactions/responses. The sounds not only reverberate in your eardrums, but also in your breastbone, literally filling your heart with music."

Most amusics will never feel confident enough in their singing to ever join a choir, even one without auditions, so they may never be able to fill their hearts with music. But that doesn't mean they can't perceive emotion in what they hear. Isabelle Peretz and her colleagues recently researched this and discovered that amusics could recognize the four emotions in the test—happiness, sadness, fear, and peacefulness—just fine. Although the subjects were indifferent to dissonance and had trouble telling major mode from minor mode, they were sensitive to tempo and timbre. "A musical emotion processing deficit is present but subtle in congenital amusia," the researchers concluded, "and can be compensated for by a normal use of tempo, pulse clarity, large mean pitch differences, and timbre in most circumstances."

But perhaps amusics aren't just compensating for what they can't hear. The Western emphasis on pitch has more to do with culture than science, and we assume that music is universal because all societies make it, but does that mean everyone hears it the same way? McGill University's Stephen McAdams and some colleagues from the Technical University of Berlin and Université de Montreal looked at whether the emotional response to music is the same across cultures by testing forty Canadians and forty Mbenzélé Pygmies. The former had no previous exposure to Congolese music and were all amateur or professional musicians. The latter live in the rainforest of northern Congo. Their isolated hunter-gatherer communities have no electricity (and no radios, television, or computers), so they had no previous exposure to Western music. Their word for music means "you can dance to it," and there are no non-musicians in their culture; everyone engages in making music, which is notable for its polyphonic complexity, for ceremonial purposes.

In the study, the subjects listened to eight selections of vocal Pygmy music that had been recorded in the field and had emotional connotations, eight Western orchestral pieces known to elicit different emotional responses, and three excerpts from film scores selected for different emotions (happiness, sadness, and fear). The participants subjectively rated what they heard on a valence scale (positive, or happy, emotion to negative, or sad, emotion) and on an arousal one (calm to excited). The researchers also measured physiological responses (including heart rate and respiration) and facial expressions.

"Is music really the universal language of the emotions? Turns out partially, but not completely," said McAdams of the results. The Mbenzélé Pygmies judged all of their music as positive and arousing, while the Canadians tended to rate it in basically the same direction, but less strongly. And some of what we hear in Western music in terms of positive and negative emotion was lost on the Congolese. They were not sensitive to mode (whether the music was in major or minor), and they didn't reliably hear the music as happy or sad.

Perhaps these results were not such a surprise, given that all of their music is upbeat. Like Westerners, they use music to modulate emotion, except they're always modulating it in the same direction. If they're sad, they play a happy song; if somebody died, they play a happy song; and if they're happy, they play a happy song. "They don't do sad songs," said McAdams. "And do you know why? Because it would perturb the harmony of the forest."

While the study found that some aspects of music were just cultural, it also showed some features of music are indeed universal. The Canadians and the Congolese had similar emotional arousal responses to faster paced music, higher frequencies, and brighter sounds — tempo, pitch, and, timbre.

There was that term again: timbre. I needed to understand it better, so I decided to visit McAdams, perhaps the world's leading expert on the subject. Originally from California, he spent twenty-three years in Paris, at the Institute for the Research and Coordination of Acoustics and Music and the French National Centre for Scientific Research. But in 2004, he returned to McGill, where he'd done an undergraduate degree in experimental psychology (and where his first experiment was on the perception of timbre). It had taken me a long time to arrange a meeting with him, so I was surprised that when I finally did get to his office late one afternoon at the end of term, he was totally relaxed, unrushed, and quick to laugh. Timbre was his favourite subject, he admitted, so he was happy to talk about it, but he also seemed cool with just chatting about music. Our conversation ranged from my upcoming house concert, which was just days away and seemed to amuse him, to *It Might Get Loud*, the documentary film about the electric guitar starring Jack White of the White Stripes, Led Zeppelin's Jimmy Page, and U2's The Edge.

McAdams studied music composition and theory before becoming a scientist. One consequence of that is that he's not interested in giving his subjects fMRIS or EEGS. "No, I don't do brains," he said. The other consequence is that he studies people's reactions to pieces of music ("real music," he called it), not to tones or tone sequences or even synthesized melodies. Unlike some of his colleagues, he studies music for music's sake, not as a path into the brain. "If you try to distill it down, I don't think you get the same impact," he said. "It's like listening to player pianos—they don't play with musical inflections."

His interest in timbre also separates him from most

other researchers. Timbre's role in music has always been underrated, or even ignored, probably because it is an intangible that's difficult to describe, hard to categorize, and so far, immune to measurement. Pitch, on the other hand, is simple to explain, fits neatly into systems such as the pentatonic scale, and can be easily measured in hertz. It's also overtly mathematical: an octave, for example, is the interval between one note and another one that's half or double the frequency. "Timbre has all kinds of dimensions, and it's much more fluid and more difficult to conceptualize unless you tie it specifically to instruments," said McAdams. "But even a given instrument has got this whole universe of timbre that it can occupy." An instrument's timbre can vary depending on other attributes such as pitch or dynamics. A clarinet, for example, has a round, luscious sound in the low registers, but it gets much more piercing in the high registers.

Even the definition of timbre is a matter of debate. Google it and you'll see that most sources describe timbre as the tone colour of an instrument. But that ignores many other attributes such as attack quality, which is how the start of the first note sounds as it moves from silent to full volume. "Tone colour would be very good for describing the difference between an English horn and an oboe and a flute," said McAdams. "But that's not going to work for the difference between an English horn and a snare drum or a harpsichord and an oboe." Should you ever get to meet him at a cocktail party and happen to ask what timbre is, he'd say: The thing that distinguishes different musical instruments. His more formal definition: Everything that distinguishes two sounds that are the same in pitch and loudness and spatial position and duration. Though, as he added, "that leaves a whole bunch of stuff."

I also asked Paul Swoger-Ruston to explain timbre to me. A former guitarist in King Apparatus, a respected Canadian third-wave ska band, he has a Ph.D. in theory and composition, teaches at Ryerson, and has composed for the Emoti-chair. Several years ago one of his pieces was part of an installation of the chairs at Toronto's Nuit Blanche, an annual all-night contemporary art festival. Swoger-Ruston is convinced timbre is a lot more important than people realize. In fact, he believes it's probably what we're most sensitive to when we listen to music. He offered two definitions for timbre. The first: those elements of a sound that distinguish it from another. The second: the waste bucket for everything we can't describe in music.

Sorry to say, but none of the definitions I heard or read struck me as a satisfying way of explaining it. My friend Chantal Braganza, who is married to a jazz musician, likes to say that if music had a smell that would be its timbre. This idea appealed to me, but I wondered if an even better way to put it might be to think of timbre as music's terroir.

In wine, terroir is the characteristic flavour shaped by where the grapes grew, including the soil, the geography, the topography, and the climate. In music, different instruments have different timbres the same way that two grapes have distinct tastes. But more than that, two guitars made by different luthiers have their own individual timbres, and two musicians playing the same guitar will have different timbres.

McAdams laughed when I suggested that timbre was music's terroir. But he didn't call it a completely stupid notion. In fact, he pointed out that there were a lot of similarities between the way we describe the taste and smell of wine and the way we describe the timbre of an instrument. Without a clear lexicon, we have to borrow vocabulary from elsewhere when we talk about these things. Oenophiles may describe a wine by comparing it to flavours such as cherry, coffee, or chocolate and some-times to something we wouldn't want to drink such as leather or charcoal. With timbre, the descriptors tend to be abstract and often borrowed from the tactile (warm, cold, rough) or visual (bright, dark) senses. The absence of a strict coding system means that some characterizations are noteworthy for their inventive evocativeness: one of McAdams's research subjects described a sound as being like someone slapping a steak on a corrugated tin roof.

The lack of respect for timbre is especially surpris-ing given that we hear it all the time. "What is speech?" McAdams said. "Speech is timbre." The unique timbre of my voice—the attack characteristics, my consonance, the colour of my sound and how it all changes over time—is based on both nature (the physiology of my head, nose, mouth, and throat) and nurture (where I carry tension in vocal production, habits of speech, and dialect). My friends recognize me immediately on the telephone or, because I have a notoriously loud voice, before they even see me in a roomful of people. It's my terroir.

"Higher Ground"

"It's a slow melodic line that climbs up with a swell in dynamics and then it comes back down again with a diminuendo." Stephen McAdams had been leaning back in his chair in a casual end-of-workday position, but he sat up a little bit as he described a musical phrase at the beginning of Richard Wagner's *Parsifal*. "He starts in with the violins but it's thickened by some bassoons and cellos and violas."

McAdams wanted to make a point about the role of timbre in the waxing and waning of tension in Western music. "As it starts to swell, he's adding in clarinets, English horn, and, at the peak, he adds in the oboes so he's making it all very much brighter. Then as it goes down, all these instruments fall out." Because he has the instruments play the same pitches, Wagner creates a powerful effect by changing the timbre. "So he's enhanced the pitch contour with a timbral contour. Actually, I am getting goosebumps just thinking about it" — he paused briefly to laugh deeply (at himself, I assumed)—"because he does it so beautifully."

Naturally, after hearing McAdams talk about this work, which I was not familiar with, I had to hear it for myself. I found it on YouTube and, even on my first listen and despite being far from an expert in opera, I immediately understood why *Parsifal* moved him so much. Goosebumps are a physiological response to music for some people, though rarely one I've experienced. But I could feel the tension build and then fall away. I felt shivers in my spine.

The power of timbre is something we are (or should be) reminded of when we go to the movies. "Why do they use these low contrabasses and bassoons," McAdams said, dropping his voice low, "in the theme for *Jaws* or those high, screechy violin sounds"—here he made high-pitched squeaks—"in *Psycho*. All that's timbral."

Another way to think about timbre is by comparing instruments. In an old stand-up bit, Steve Martin strummed the banjo and mused, "You just can't sing a depressing song when you're playing the banjo . . . You can't just go, 'Oh, death, and grief, and sorrow, and murder.' . . . When you're playing the banjo, everything's okay . . . I always thought the banjo was the one thing that could have saved Nixon." And in "You Just Can't Play a Sad Song on a Banjo," Willie Nelson sang that violins are sweet and steel guitars are thrilling, but nothing beats banjos for fun.

Banjos have bright timbres and the plucked strings have a short sustain, which encourages musicians to play them fast, according to Ohio State University's David Huron. You aren't likely to hear sad songs played on piccolos either, because of their high pitch and bright sound. Musicians seeking the melancholy might want to try a cello. Because of its low range and sustained pitches, it can be played slowly and quietly. Huron's study found that part of the reason we may consider some instruments sad is that they have some of the same attributes — quieter, slower, with smaller pitch movements, relatively lower pitch, more "mumble-like" articulation, and darker timbre — that convey sadness in speech.

Some instruments generate different emotions depending on how they're played. Brass instruments deliver a much brighter, brassier sound when played loud than when played softly. "One of the powerful things about a brass crescendo," said McAdams, "is it's not just getting louder, it's also getting much brighter at the same time. So that's why it's a really effective emotional device in orchestral music."

Gamelan is traditional Indonesian music played by ensembles of metallophones, drums, gongs, bamboo flutes, and other instruments. I'd never heard of it until I took the first of the two classical music courses, which included a World Music component. Swoger-Ruston was fascinated when I told him it was my favourite music from the courses because Gamelan is all about timbre with little in the way of pitch movement.

I checked the textbook to see what it had to say about timbre. Not much. According to the index in *Understanding Music*, the word appears on two pages and is defined as a synonym for tone colour, which appears on a few more pages. But timbre has deep roots in Western classical music. During the Middle Ages and Renaissance, composers didn't specify what instruments should play the music. It was all about pitch, and it didn't really matter what instruments the musicians used—they just performed with what they had. But starting in the Baroque period, some composers, notably Bach and Vivaldi, began to specify the instrumentation as they sought certain sounds and experimented with different combinations of them. Timbre, in other words. Vivaldi's *The Four Seasons*, for example, used specific string instruments to convey the sounds of the seasons and nature.

This playing around with the possibilities came at a time when instrument makers were doing some of their own inventing, giving the composers more options. Later, Haydn, Mozart, and Beethoven worked with orchestration, and in the latter half of the nineteenth century the size of orchestras grew and composers became increasingly concerned with what they could get out of each instrument. Meanwhile, the artisans who made instruments kept refining their products to elicit new and different sounds. At the turn of the twentieth century, Claude Debussy famously worked with timbre. Many people called his compositions Impressionism because they seemed to be the aural equivalent of what the Impressionist painters were doing (though, predictably, he hated the term). After that came Edgard Varèse and his "organized sound," Arnold Schoenberg and serialism, minimalism, and the French spectral school. According to Swoger-Ruston, "The twentieth century was really about explorations in timbre."

As venues changed, so did music. In the Middle Ages,

musicians often performed in Gothic cathedrals and other stone-walled places as part of religious functions. "The reverberation time in those spaces is very long—more than four seconds in most cases—so a note sung a few seconds ago hangs in the air and becomes part of the present sonic landscape," writes David Byrne in *How Music Works*. "A composition with shifting musical keys would inevitably invite dissonance as notes overlapped and clashed—a real sonic pileup." Several distinct timbres would have created their own sonic smash-up. So Gregorian chant, which was monophonic—a single melodic line with no harmony or countermelody—also required the singers to match timbres. Music became more complex as it moved to concert halls, opera houses, and palace ballrooms. Larger orchestras may have been a result of the need to resonate more in ballrooms full of dancers. And as music became louder, it allowed for more variety in timbres.

If classical composers explored timbre, popular music, which is often played in even less controlled environments, became all about timbre. The high nasally voice that we hear in early country and early blues is probably a function of playing in noisy juke joints without microphones. "Those tones really cut through the audience," said Swoger-Ruston. "So shaping your voice for projection ends up creating a timbral quality to that music." Similarly, as rock stars began playing large venues such as hockey arenas, the subtleties of harmony, if there were any, became lost. Byrne says the sound in CBGB, the legendary New York City club where he both listened to and played music, was remarkably good. "The amount of crap scattered everywhere, the furniture, the bar, the crooked uneven walls and looming ceiling made for both great sound absorption and uneven sound reflections—qualities one might spend a fortune to recreate in a recording studio." And with no reverberation, the audience could hear every detail of the music, as long as it was played loud enough to overwhelm the unruly crowd. Think about that the next time you listen to "Psycho Killer," the first Talking Heads hit.

Timbre is even more crucial in contemporary genres such as electronic dance music and rap. EDM, said Swoger-Ruston, is really "experimentation with timbre." The widespread use of sampling means the sample, not the note, is increasingly the basic unit of music.

Ever since nineteenth-century Romanticism, we've had this idea that the musician is supposed to be an individual. Guitarists create their own unique sound depending, among other things, on where they're plucking the strings and at what angle. And they are increasingly playing with effects boxes and other gadgets. One of McAdams's favourite guitarists, The Edge, is forever changing the timbre of his sound with such technology. "That's a whole timbral universe that he's trying to use that increases the emotional impact of what he's doing," said McAdams. "So he's controlling pitch, he's also controlling timbre a little bit with the way he plucks, but he's controlling timbre a lot with all of his effects boxes and things like that."

Nothing is as distinctive as the human voice, so it's no surprise that vocal timbres would become so pivotal in popular genres. Folk music, for example, is really about the individuality of the voice, often accompanied by guitar playing that is little more than functional and supportive. "Certainly just having a stripped down texture of guitar and voice is a unique timbral condition," said Swoger-Ruston. He also believes that 1960s rock is built on intense layers of texture and timbre almost to the point of saturation. But once again, the vocal timbre is crucial. There aren't too many big singers from that genre that we can't identify quickly from their voice, he said, adding that a lot of that music relies on similar tools — the same chords and harmonies — and only small differences in melodies. So timbre is the easiest way to create something unique. Like me, he's usually not as attracted to virtuosic singers. "It's these voices with real character to them and maybe you hear the real person behind them."

A vocal timbre can also reveal economic and cultural background. Some singers, especially in the early periods of commercial popular music, were encouraged to either accentuate their timbre — including accent, diction, ornamentation, and register — or to abandon what occurred naturally for them, in the interest of appealing to a wider or different audience. In the late '50s and early '60s, Patsy Cline weakened her accent and strengthened her vibrato to appeal to new listeners as the Nashville Sound gained fans. According to Gillian Turnbull, these changes may have made the singer (and, by extension, the genre) more "authentic" in terms of representing a particular class of people.

The relationship between timbre and class is also a factor in the splintering of musical culture that began in the '50s and accelerated in the following decades. "Musical subcultures exist because our guts tell us certain kinds of music are for certain kinds of people," writes Carl Wilson in *Let's Talk about Love: A Journey to the End of Taste.* "It's most blatant in the identity war that is high school, but music never stops being a badge of recognition. And in the offhand rhetoric of dismissal— 'teenybopper pap,' 'only hippies like that band,' 'sounds like music for date rapists'— we bar the doors of the clubs we don't want to claim us as members."

But here's the thing: since most of the genres we fight over are built on the structures, harmonies, and basic forms of early popular music, the only way to distinguish, say, country rock from honky-tonk, or psychedelic rock from heavy metal, is by timbre. "Otherwise," said Turnbull, "how would Led Zeppelin be different from Jimi Hendrix or James Brown different from Otis Redding?" This diversity in timbre coincides with the fracturing of the popular music audience— and the vehemence with which listeners defend their favourite genres. "I think that's all based in emotional reactions to timbre before any consideration of lyrical themes or other musical elements."

A common misconception is that timbres are just carrying melodies and rhythms, but instrumentation can alter our perception of harmony. Change the instruments and the piece of music is not going to sound the same. It can sound more or less dissonant, for instance, and that ebb and flow of dissonance and consonance is the underlying experience of harmony. Timbre carries a lot of that information and we process it surprisingly quickly, which is why we can sometimes recognize a song in as little as a quarter of a second, long before we even hear the melody.

And still we don't talk much about timbre. "If you look at music theory, they talk about pitch and harmony, they'll talk about things like rhythm and meter," McAdams said. "After that it gets into things like voice leading and counterpoint and harmony. And they never, ever talk about timbre."

If someone could find a way to measure timbre, we'd likely take it more seriously. And McAdams is working on just that. It's not easy because timbre has so many dimensions. In one study, his lab analyzed over six thousand sounds from a musical database and found around ten different classes of acoustic parameters that are independent of each other. Using that model, there might be as many as ten dimensions to timbre. Capturing it mathematically will require quantifying attributes such as brightness, attack time, effective duration and noisiness. Then it might be possible to come up with the global measure, but McAdams has a lot more experiments to do before he gets there.

And that might not even be the best approach. Another possibility is that we aren't perceiving music in terms of dimensions but as a complex spectromorphology, an electroacoustic term for the sonic footprint of a sound spectrum as it changes over time. With a trumpet, for example, there's a little high-frequency energy with the attack, and then primarily the low frequencies come out first and the high frequencies come out later. And when the sound goes away, the high frequencies go first and then the low frequencies. Capturing all that information mathematically would be a difficult challenge indeed.

If measuring the timbre of instruments will be hard, imagine trying to do it for human voices. That would take us into snowflake territory (one reason McAdams has done little research into vocal timbres). We don't fully understand how people react to different vocal timbres, whereas we do know how most listeners respond to a beat that's above a resting heart rate or to syncopated rhythms or big jumps in melodies. "We have simple ways to describe where the emotional content sits in those parameters," said Swoger-Ruston. "But with the human voice it's complex and so many things are going on."

Even if McAdams can't solve the measurement puzzle, maybe his work will help more people realize that the better we understand timbre, the better we will understand music — and how we really perceive it.

The more I learned about it, the more convinced I became that timbre was crucial to what I was really hearing when I listen to music. I was also sure that I was attracted to vocal timbres more than lyrics. What I didn't know was if everyone else heard music the way I do. Being self-absorbed, and unable to help myself, I liked to ask the scientists who studied me, "So if I recommended a band to you, would you dismiss the suggestion because I am tone deaf?" They were invariably polite and said they wouldn't. Taking Frank Russo at his word, I suggested we go to a concert together sometime and then invited him to join me at a Garland Jeffreys show.

I've been a huge fan since *Ghost Writer*, Jeffreys's 1977

masterpiece that mixes rock, reggae, and a few other genres with intelligent, culturally aware lyrics. Russo didn't know Jeffreys's stuff, so I burned a CD of some of my favourite songs. He liked what he heard and agreed to go. Now in his seventies, Jeffreys may have slowed down a bit, but he still puts on a rousing performance. That night he played most of the songs that I'd included on the CD. Russo seemed to enjoy the show a great deal, which—foolishly, I'm sure— I took as a small triumph.

When I'd originally asked him whether he'd trust my tastes, Russo had mused that since he had a sense of what aspects of music I could perceive, and which ones I couldn't, an interesting project would be to look at what music I like and don't like. That might help us better understand what I'm getting out of music. But I knew by then that Russo had far more ideas for projects than time and resources to do them.

A year or so later, Turnbull and Swoger-Ruston suggested something along the same lines. Both were convinced that everyone—not just me—mostly responds to timbre. They wanted to design a study that could show that. Would I be their first test subject as they figured out how such an experiment might work? I agreed, though not before extracting a promise that when they asked what I was hearing in a song, I could also ask them what they were hearing.

We had the first session at my house. Turnbull had created a playlist that included some songs from the Top 50 list I'd created for her and some I might have different reactions to. We hooked her phone up to my stereo, and she hit play on the first song. After a while, she hit pause and asked, "The first question is just about your overall reaction. Do you like or dislike it? How does it make you feel? Do you connect to it emotionally? Does it remind you of a period in your life or something positive or negative?"

"I didn't like it," I said. "I didn't like the voice. Even before the singing—the screeching—started, I found it kinda boring. I realize that's not a technical term. I recognize who it is, but can't remember who it is. But it just strikes me as boring metalhead music."

Turnbull told me the song was AC/DC's "What Do You Do for Money Honey" and asked, "What else do you hear musically?"

"The vocal timbre is very screechy. It's got a strong beat, but kind of a boring beat. It doesn't make me want to dance, but I might bang my head against a wall."

I asked her what she liked about it. "I like the guitar sound more than anything and the rhythmic interaction between it and the drums," she said. "The voice—I can take or leave his voice."

The next selection was Calexico's "Sunken Waltz," a song from my list. "I love this song," I said. "I love the sound. It's got a warm timbre, it's interesting musically, I like the voice, it's very warm." I had trouble identifying more than a few instruments, but said, "Everything really fits together. The vocals really work with the sound of the music."

Swoger-Ruston asked what I thought of the rhythm. "It seems more interesting to me," I said. "Is lilting a bit too strong?"

He noted the waltz meter and said, "It's got this slow feel but the activity is actually much busier."

Turnbull played the part where all the other instruments drop out, leaving just the guitar. She asked, "Do you like that moment in the song?"

"I love it," I said. "It grabs my attention every time I listen to it."

She noted how gratifying it was when the other instruments come back and the lush texture returns.

After the first three songs, I pointed out that we hadn't talked about the importance of pitch in any of them. "I never think about pitch," said Turnbull. "No, I shouldn't say that. There are songs with interesting chord changes that grab me. That's about it for me and pitch."

"Well, if there's a strong melody," said Swoger-Ruston. "If there's a really interesting turn in the melody, then I will focus on it."

And as it turned out, the subject came up on "Car Wheels on a Gravel Road" by Lucinda Williams. "I think pitch matters in this one," Turnbull said. "She's basically dancing around about three pitches there, just staying on the same note and going slightly down and slightly up — except for the one leap she has. In the verse, the bass is really interesting. It's playing a melody that's more interesting than her vocals and then you get to the chorus and her melody becomes more interesting — she starts singing higher and has a bit more movement going on — and the bass gets super boring. It just does the bass thing." But I had trouble picking out the bass, because I was getting distracted by the higher-pitched mandolin, until Turnbull played the track again and Swoger-Ruston sang the bass part.

Next up: "Higher Ground." But not the Stevie Wonder original; instead, Turnbull had picked a funk-filled cover by the Red Hot Chili Peppers. "It has a beat that's fun," I said. "And it seems like a lot's going on. You could work out to this. And you definitely could go driving. Fast." I asked Turnbull why she'd played it for me. The timbre, she said, the sound of the bass in particular. "It gets me right away."

"Bright guitars, fat bass," said Swoger-Ruston. "A lot of people would find that texture overwhelming because it's so much information. So I think it's significant or at least interesting that you like that kind of complexity in that music."

"But it aligns with his response to, say, Calexico," said Turnbull. "There's a lot going on there, too, but it's all joining together for an overall effect."

"Here it's just kind of extreme," he said.

"Yeah," I agreed. "It's *bam*, in your face. But in a fun way."

The final song was one I adore: Neko Case's "Thrice All American." I jumped in without waiting for Turnbull to ask a question: "Well, the voice. The lyrics. I guess musically it's not all that interesting. Her later stuff is more interesting musically. But it starts off and her voice is the main thing and then more instruments come in, but it's still the voice."

"What do you like about her voice?" Turnbull asked. "That it's technically beautiful or does she get you somehow emotionally with the way she sings?"

"I would say both. It's a very evocative voice. But even as

someone who doesn't hear pitch well, I can tell she's doing great things with her voice."

We discussed the song's triple meter and Turnbull said, "In talking about these songs, it's become clear that you like complex arrangements."

"Yeah," Swoger-Ruston agreed, "you like continuous change."

I was happy that I had not revealed myself to be a tone-deaf fool. Neither Swoger-Ruston nor Turnbull had said, You're not hearing that? or You're hearing what? And none of their comments made me think I was missing too much, though they had more educated ears and knew the musical terms so they could talk about songs in deeper, more specific ways. I wasn't great at picking out all the different instruments in a song, but they both agreed that probably had more to do with training than amusia. (Turnbull had only asked me about what instruments I was hearing to see if there were certain timbres I liked more than others.) In fact, Swoger-Ruston thought I was hearing what other engaged listeners hear. "It's like talking to someone else who loves music a lot and pays attention to it."

That was an encouraging bit of validation, but I was still a guy swayed by science. And Russo was a scientist who had studied me, followed my progress as I tried to learn to sing, and had many conversations with me. So it made sense for me to ask him one last time what he thought I was hearing in music. He didn't think it was a matter of me missing something so much as a case of me giving different weight to the various dimensions of music, including melody, harmony, rhythm, timbre, and so on. How much or how little our brains value each of these dimensions may be a result of culture—listening history, for example—as ethnomusicologists would have it, or of congenital or acquired neurological predispositions. So if you took ten people off the street, they'd all have a different formula for how they listen to music.

For me, timbre is probably the most powerful. And rhythm might be important, too. That doesn't mean that harmony and melody don't matter at all to me, just that timbre means more. For a musically trained person or someone who studies music in university, harmony might be at the forefront. The average listener may be somewhere in between on harmony. Or, if Turnbull and Swoger-Ruston were right, maybe the average listener was closer to me on timbre. Regardless, I liked the idea that we all have our own recipe for loving music.

Even better, Russo added the most intriguing and (perhaps unintentionally) complimentary thing anyone could say to an amusic person, at least one who loves music. "If you were to offer a music course, I think it could generate insight for regular auditory listeners," he said. "We'd learn from your experience, your way of understanding things."

I would relish a chance to teach such a course. I'd call it Unlocking a Surprising Secret of Music and I'd tell my students that they are hearing a lot more timbre than they realize. Maybe not as much as I do, but more than they have been taught to believe they do. So they should never underestimate timbre, especially since that's where so much of the emotional power of music comes from. But mostly, I would tell them that my experience of trying to learn how to sing and understand what I hear has taught me what perhaps should have been obvious all along: There is no wrong way to listen to music.

"Silver and Gold"

Florence Foster Jenkins made it to Carnegie Hall when she was seventy-six. If her age didn't make the concert unusual, the American soprano's complete lack of singing ability sure did. She had no sense of pitch or rhythm, and her voice had a tendency to disappear on high notes. And yet she held annual recitals at New York's Ritz-Carlton Hotel until she booked Carnegie in 1944. "She was exceedingly happy in her work," wrote Robert Bager in his *New York World-Telegram* review of the sold-out show. "It is a pity so few artists are. And her happiness was communicated as if by magic to her listeners... who were stimulated to the point of audible cheering, even joyous laughter and ecstasy by the inimitable singing." For her part, Jenkins, who died shortly after the infamous gig, said, "People may say I can't sing, but no one can ever say I didn't sing."

I will never make it to Carnegie Hall, just as I'll never play right wing for the Boston Bruins or any other team in the National Hockey League. I'm okay with both of those disappointments, but I still play three hours of hockey a week and get a lot of delight out of it. All I want is to be able to make music as badly and enjoyably as I play old-timers hockey.

And so in the spring of 2015, I set a date for my perform-ance. Finally, Barnes and I had a deadline. This focused my mind in a way it hadn't been before, and Barnes noticed that I was practising even more. I'd long ago given up wait-ing for something to just click and suddenly I'd be able to sing in tune. Now I believed that if I worked hard enough to learn the contour of a song and received enough feed-back from Barnes, I could develop the muscle memory to produce a not-unacceptable rendition. And it seemed to be working. I was getting better. Even if I couldn't be entirely sure I was hitting the notes, because I couldn't hear well enough, I sometimes knew when I hadn't done a good job because it didn't feel right.

In my mind, at least, the song I was going to sing was "Silver and Gold," but one day Barnes suggested I learn "Blackbird." It would be a much tougher song to sing, but by this point, he assured me, I was up for the challenge. I couldn't help but think that, given how that young woman had mocked me for hitting a right note in that song all those years ago in Vermont, singing it decades later would truly be an ironic performance. But I kept that baggage to myself and didn't tell Barnes.

Meanwhile, he was working on getting a better sound out of me. If I wasn't going to hit all the notes, at least I could get some of the other things right so I sounded more like a singer. And if I stopped worrying about hitting the notes perfectly, then I might have a better chance to get the other stuff right. My breathing was starting to improve, though there were still times when I would forget to breathe. (Yoga instructors have told me the same thing when I concentrate more on a weird pose than on my breath.) And my sound lived in the wrong place, so I needed to move it from back in my throat to the front of my face where it could resonate in my skull. Barnes asked me to hold a finger from each hand to the sides of my nose and told me to sing to my hands. One more thing to worry about.

My gig was three weeks away when we stumbled on a lyric change that we both thought would be perfect for my performance of "Silver and Gold." I'd printed out the lyrics; I knew them but found it handy to have the piece of paper for when I lost my place. Barnes had asked for the sheet and when I stumbled, he fed me the line: "Oh, I do a lot of things I know is wrong / Hope I'm forgiven before I'm gone." When I started to giggle about my inability to remember a line I knew so well, he sang, "I sing a lot of notes I know is wrong." We both cracked up.

"I should sing that," I said, thinking it might help break the tension for any of my friends who would be feeling nervous for me.

"You totally should," Barnes said. "People will love you for it and you will get a big laugh."

And a laugh with me would be much better than a laugh at me, which was what I feared the most.

We practised the new lyric, chuckling each time, before working on "Blackbird." As I was leaving, Barnes said he was really happy with my progress. He assured me he didn't think my performance would be a disaster. I'd be nervous, but I'd have fun. My "Blackbird" was in really good shape, he told me.

"Better than 'Silver and Gold'?" I asked.

Apparently. While the latter was a simpler song, I was talking it too much, instead of singing it, especially the last line of each verse. Strummer talk-sings it, too, but without losing the melody. The advantage of "Blackbird" was that I couldn't cheat and talk-sing it, and somehow I was doing pretty well with the melody. In fact, instead of being my back-up song or the song I'd sing if my friends insisted on an encore, Barnes was suggesting that was the song I should sing.

How could it be possible that I could sing a Beatles song better than one by The Clash's Saint Joe Strummer? Many of my friends think I hate The Beatles. After reading a magazine article I'd written about my tone deafness, a friend emailed to say: "At least we finally know why you don't like The Beatles. You're hearing something else." I don't really hate the Fab Four, of course. I'd loved them as a kid, but after I started listening to The Clash, they suddenly seemed overhyped, overplayed, and old-fashioned. And it is annoying to hear fans go on and on about them as if they were the only band that ever mattered. Which is obviously factually inaccurate. So I don't own any Beatles albums; buying them seemed redundant anyway, since I hear the songs everywhere from car commercials to store sound systems. (It's like a slightly hipper version of Muzak.) I've actually always liked "Blackbird," but picking it for my performance would be a completely different kind of musical joke than our lyric tweak—though, I suppose, not one that as many people would get.

In one of the most charming myths of popular music, Robert Johnson went down to the crossroads of two Mississippi highways, sold his soul to the devil, and returned an incredible bluesman. That's a deal I'd gladly make just to sing in tune. No such pact was ever going to happen, but I knew I could never call myself a singer if I couldn't do it in front of other people. So my attitude was: I'm going to sing and I'm sorry.

At a session a couple of weeks before my gig, as Barnes and I were jokingly calling it, he asked if I was getting nervous yet. If I was having trouble sleeping, for example, we should talk, he said. But I was fine. And I didn't really start to think about what could go wrong until a few days before the house concert. I reasoned that I'd lowered everyone's expectations. As long as I didn't piss my pants and run out of the house, I would be fine.

My friends Kelly Crowe and Paul Hamel had agreed to host. They had the perfect house for the event—they'd recently completed a renovation that had turned the main floor, which had been several small rooms, into one large gorgeous space. Even better, they had a P. A. Starck baby grand piano. I sent an invitation to a few friends with the subject line "Cone-of-silence house concert." I explained that I was doing a secret gig and why. "I would like you to come if you promise you won't laugh too hard. Here's the deal: come at 9; I will sing at around 9:30; then you're invited to get up and do a song or two (so bring your instrument if you're interested in showing me up); then around midnight, we'll turn the mic off and have a vinyl dance party. My one request: please do not talk about this with anyone — I'm under enough pressure as it is."

As the night of the party approached, Barnes added one more thing to think about. Just sing it emotionally, he urged me. Don't worry so much about the notes, just try to feel the song, what it means to you. We'd talked about this before, but now he was insistent that I find a way to connect to my material. That was going to be easy with "Silver and Gold," which I already related to. But I didn't have an emotional bond with "Blackbird," so Barnes suggested I go home and spend some time with the lyrics.

The house concert was on a Saturday, and I arranged vocal sessions for Thursday afternoon, Friday morning, and Saturday afternoon. I didn't sing that well on Friday and Barnes could tell I was nervous, though I felt okay. The Saturday session seemed better. Afterwards, I rode my bike home and ate dinner, then shaved and showered. I had no performance clothes — one friend said he was surprised I hadn't worn a tux — but I put on black wool pants and a loose-fitting button-down collared shirt. I was used to sucking my gut in when I was in public, but if I was to breathe properly, my belly would stick out. That meant I needed to wear a big shirt, even if it didn't look great.

Barnes had generously offered to play with me. Since I

wasn't sure when to come in and I had trouble always singing the words at the right times, he was willing and able
to follow me and speed up or slow down when necessary.
We met at the venue half an hour early so we could do
a final run-through while our hosts did some last-minute
setting up. The place looked great—they'd even placed a
candelabra on the piano to evoke just the right amount of
Liberace. The guests started to arrive right at nine, and I
endured half an hour of going through the motions of greet-
ing people. My mind was elsewhere, and after I said hello,
I had nothing more to say. I was too distracted and mostly
stood by myself as the guests grabbed drinks and chatted
with each other. One friend asked if I'd taken beta block-
ers to calm my performance anxiety (I hadn't). Another
pointed out that I was wringing my hands, which was weird
behaviour for me.

By 9:35, about forty-five people were there. I knew most
of them, but Kelly and Paul had also invited a few strangers,
increasing the degree of danger. I took my spot beside the
piano, facing the crowd. I had the lyrics on a music stand
in front of me, but I could still see Barnes. After thanking
everyone for coming, I explained that I was part of the
2.5 percent of the population that's scientifically tone deaf.
But I was going to sing anyway.

I was really nervous now. Or not so much nervous as
completely, but not comfortably, numb. My face reddened
and I kept running my hand through what's left of my hair,
which is a nervous tic I have. But I didn't experience stage
fright. I really did want to get up there and sing. Perhaps
years of teaching classes and doing readings and other

public speaking engagements gave me some confidence to stand in front of an audience. Plus, I've always craved attention—you try being the only boy in a family of five kids. Sure, I was worried I wouldn't do a good job, but calling it off never crossed my mind.

And as soon as I started singing "Blackbird," nothing crossed my mind. Everything went out of my head: the breathing, the singing to the nose instead of back in my throat, the idea of communicating, and of course, the notes I was supposed to be hitting. Although I knew I had sung the song better many times in Barnes's studio—including that afternoon—now I was just under too much pressure. Barnes later emailed to say, "Hey, it's par for the course to be too freaked to remember much technique the first time (sounds like other situations we have known :) BUT amazingly the hours of hard work paid off and your singing Sat night was indeed what people recognize and describe as 'singing'!"

My biggest regret: I'd set the music stand low because I hoped that I would need to glance at the lyrics only occasionally. But I ended up staring at the words so my eyes were down almost the whole time. Still, when it was over, the affectionate applause from my friends was a bit overwhelming.

The reviews after my performance were that "Blackbird" was pretty rocky, especially at the beginning. I'd started off on the wrong note, and it wasn't until the second verse that I started to recover and then only intermittently. I couldn't hear that I had begun badly, but I knew because it didn't feel right. A few lessons earlier, Barnes had suggested that if I started on the wrong note, he would stop playing and we could begin again. But I wasn't keen on that idea. I wanted to keep going no matter what. And that's what I did.

Fortunately, I couldn't see or hear my friend Biff. As soon as I started singing, he fell into a giggling fit and had to run out the back door. (Later, he told me, "No one will doubt the premise of your book.") Nor did I have any idea how nervous Carmen was or that she was trying to will me through it.

After the applause for "Blackbird," I started singing "Silver and Gold." It was the easier song and the one I related to more, so I was much more into it. I tried to have fun and even looked up a couple of times. The crowd was less tense for me, too, and the new line about singing wrong notes earned some appropriate laughter. When I finished, there was more enthusiastic applause and even calls for an encore. But I'd prepared only two songs and I had just performed them. If I had learned anything from this experience, it was that trying to improvise was a bad idea indeed. I needed to practise the hell out of a song before I could even consider doing it in front of people.

I was relieved it was over, but also pumped that I'd done it. At some level, belting out two songs in front of a roomful of people was a rush (and, yes, the terror was definitely half the thrill). Even if it wasn't an accomplished performance, I had—for a few minutes, anyway—been the lead singer.

I joined the crowd and Barnes played a song from his album *New York Stories*, which was to come out ten days later. He also took the opportunity to tell my friends that he was honoured to be part of my journey and to say how hard I worked. Then he looked at me and said, "You are a hard-working motherfucker."

Afterwards, as I expected, I heard a couple of wry comments about singing a Beatles tune despite owning no Beatles albums. But the word I heard the most was brave. I knew that one person's brave is another person's foolhardy, but I didn't really care: I had sung in public. And no matter how bad a singer I was, I was better than when I started.

Later, my regular concert date Bill Reynolds said that hearing me sing helped him understand my disorder a little better. He was fascinated that, while I was off the whole first verse, I was consistently off. That made him realize that my training had helped me be able to control my voice, but I still couldn't tell if I was on pitch or not, which made my condition all the more poignant. Paul Swoger-Ruston noted that there were sections where I was clearly off pitch in both songs. But, he told me, "then there would be whole phrases where you were just right locked in and tuned." Maybe I didn't hear when that was happening, he said, but I must have felt it because when I locked in, all of sudden I looked relaxed and my voice became much stronger.

My singing career had lasted all of two songs. Like the professional athlete who gets the proverbial cup of coffee in the bigs, I wished it had been longer, but I was proud to have been there at all. Although I hadn't been able to cure my amusia with my lessons, I'd improved enough to make me wonder what would have been possible if I'd started earlier and kept at it longer. And if anything, the whole experience had made me love music even more.

After several of the guests took turns playing songs, I

passed around the lyrics to "I Shall Be Released," and we all sang along as if we were part of the finale of *The Last Waltz*. We didn't sound as fabulous as The Band and friends had at that Thanksgiving concert in 1976, of course, but there really was something charming and uplifting about singing together. Good singers, bad singers, it didn't matter — we all belted it out. More than just connected, we were all joyful.

And then we danced.

ENDNOTES

Epigraph
How Music Works by David Byrne (McSweeney's, 2012),
p. 155.

Side One
Track Two
Generation Jones refers to people born in the second half of
the demographic baby boom (the mid-1950s to the mid-'60s).
Although lazy journalists and pundits often lump these
people together with Baby Boomers, this is a distinct cohort
with attitudes and values shaped by different historical,
cultural, and economic events (Watergate instead of the
Vietnam War; punk rock instead of the Beatles; the energy
crisis, stagflation, and the Reagan recession instead of the
expansionary '60s). Alas, the term, coined by social com-
mentator Jonathan Pontell, hasn't caught on with the general public, though it certainly deserves to be common coin.

The stats on listening habits are from "Generation M2: Media in the Lives of 8- to 18-Year-Olds" (Kaiser Family Foundation, 2010).

"Active listening" is from *Understanding Music,* 5th ed., by Jeremy Yudkin (Pearson Prentice Hall, 2008), p. 62.

For all their convenience, YouTube and streaming services are changing our relationship to music, and not necessarily for the better. The treasured album, purchased with care and cash, was special because it was rare: even hardcore collectors were choosy about what they bought. Now we can listen to anything anytime and it no longer seems so special, though it does mean we're willing to try more artists and more genres and, if anything, we're listening to more of it rather than less.

Sasha Frere-Jones's article on Brian Eno is "Ambient Genius," *The New Yorker,* July 7, 2014.

For more on specific musical anhedonia, see "Dissociation between Musical and Monetary Reward Responses in Specific Musical Anhedonia," by Ernest Mas-Herrero, Robert J. Zatorre, Antoni Rodriguez-Fornells, Josep Marco-Pallarés (*Current Biology,* 2014).

For more on the flutes found in Germany, see "New Flutes Document the Earliest Musical Tradition in Southwestern Germany," by Nicholas J. Conrad, Maria Malina, and Susanne C. Münzel (*Nature,* 2009). The research arguing that the Divje Babe flute is not, in fact, a flute is in "'Neanderthal Bone Flutes': Simply Products of Ice Age Spotted Hyena Scavenging Activities on Cave Bear Cubs in European Cave Bear Dens," by Cajus G. Diedrich (*Royal Society Open Science*, 2015).

How the Mind Works by Steven Pinker (W. W. Norton and Company, 1997). I quote from pages 528 to 538.

Dan Sperber calls music "an evolutionary parasite" in *Explaining Culture: A Naturalistic Approach* (Wiley-Blackwell, 1996).

Guitar Zero: The New Musician and the Science of Learning by Gary Marcus (The Penguin Press, 2012). Puzzling quote is from page 5. His point about the advantages of a good sense of pitch is from page 29. The quote about *"cultural* selection" (his italics) is from page 118.

Levitin on music as an evolutionary adaptation: *This Is Your Brain on Music: The Science of Human Obsession* by Daniel J. Levitin (Penguin, 2006), pp. 249 to 253.

"More Than Cheesecake?" is the second chapter of *The Singing Neanderthals: The Origins of Music, Language, Mind, and Body* by Steven Mithen (Harvard University Press, 2005), pp. 11–27.

The Darwin quote is from *The Descent of Man, and Selection in Relation to Sex* by Charles Darwin (1871).

The study on how Neanderthals and other human ancestors used a music-like signal is "Did Neanderthals and Other Early Humans Sing? Seeking the Biological Roots of Music in the Territorial Advertisements of Primates, Lions, Hyenas, and Wolves" by Edward H. Hagen and Peter Hammerstein (*Musicae Scientiae*, special issue 2009–2010).

The story about what happened during the performance of Handel's *Messiah* at the Bristol Old Vic is from "Leading Scientist Ejected by Audience after 'Trying to Crowd Surf' at Classical Music Concert." *The Independent*, June 20, 2014; and "At Handel's Messiah, He Was a Very Naughty Boy! Leading Scientist Kicked out of Classical Music Concert after Trying to Crowd Surf," by Emma Glanfield (*The Daily Mail*, June 22, 2014).

Parts of the section on Tom Chau and the VMI originally appeared, in a slightly different form, in the Canadian edition of *Reader's Digest* as "Techno Music" by Tim Falconer (August 2012).

Steven Brown also proposed a precursor to both music and language in "The 'Musilanguage' Model of Music Evolution," a chapter in *The Origins of Music* (MIT Press, 2001).

Track Three

The Lyle Lovett story is from "Lyle Lovett: Home Is Where the Art Is" by Richard Ouzounian (*Toronto Star*, August 6, 2010).

The parenting book was *Drop the Worry Ball: How to Parent in the Age of Entitlement* by Dr. Alex Russell with Tim Falconer (John Wiley & Sons Canada, 2012).

Further to my point about the professionalization of music: Sure, popular musical acts inspired lots of people to pick up guitars and other instruments. But a moody teenager sitting alone in a bedroom strumming a guitar—even if he or she ends up a rock star—isn't the same as a family singing together around the piano in the living room.

All religions feature hymns or chants in their rituals and services. The Ancient Greeks summoned their gods with song and even had a god of music (Apollo). Starting in the Middle Ages, the Roman Catholic Church used Gregorian chant in its services. Hindus sing devotional songs called bhajans. And so on. Some of the greatest classical music ever composed, including Palestrina's *Pope Marcellus Mass*, Bach's *St. Matthew Passion,* and Mozart's *Requiem,* was written for the church. And even as an atheist, I have a soft spot for Billy Bragg's version of "Jerusalem." Whenever people get together—weddings, parties, anything (including funerals)—there's going to be music, though increasingly professional players or recordings provide the music where once the celebrants created it.

Byrne on the professionalization of music: *How Music Works,* pp. 271 and 279.

The famous, though frequently loosely paraphrased, Brian Eno quote about the first Velvet Underground album inspiring so many bands is from "Eno: Voyages in Time and Perception" by Kristine McKenna (*Musician*, October 1982).

Track Four

Pfordresher's teacher was Phil Tacka, then at Georgetown and now at Millersville University of Pennsylvania.

The bit about Luciano Pavarotti's rich tenor not being right for country music and Johnny Cash's deep growl being ill-suited for opera is based on something Sean Hutchins said to me.

The story about Neil Young is taken from *Tears Are Not Enough*, a documentary about the recording of the song (https://youtu.be/AhfDW2pqros).

For more on our perception of the accuracy of operatic singing, see: "Criteria and Tools for Objectively Analysing the Vocal Accuracy of a Popular Song" by Pauline Larrouy-Maestri and Dominique Morsomme (*Logopedics Phoniatrics Vocology*, 2014).

For more on the reasons some people are poor-pitch singers, see: "A Frog in Your Throat or in Your Ear? Searching for the Causes of Poor Singing" by Sean Hutchins and Isabelle Peretz (*Journal of Experimental Psychology*, 2012).

The distinction between how Hutchins and Pfordresher measure bad singing largely hinges on the difference between consistency or overall accuracy. Hutchins used a measure that was sensitive to both consistency and accuracy and under that criteria, someone who sings the correct notes on average, but not consistently so, could be considered an inaccurate singer. The Pfordresher and Brown study was looking for people who were consistently sharp or flat. "I would not go so far as to claim that one measure is always better than the other," Pfordresher, who has used both, explained in an email. "They tell us different things about singing accuracy. My analyses suggest that people who are inaccurate in the Pfordresher and Brown sense probably have a more serious problem than those who are inconsistent."

The Pfordresher and Brown study is here: "Poor-Pitch Singing in the Absence of 'Tone Deafness'" by Peter Q. Pfordresher and Steven Brown (*Music Perception*, 2007).

For more on why we are more forgiving of singers than violinists, see: "The Vocal Generosity Effect: How Bad Can Your Singing Be?" by Sean Hutchins, Catherine Roquet, and Isabelle Peretz (*Music Perception*, 2012).

Side Two
Track Five
You can take the BRAMS online amusia test here: www. brams.umontreal.ca/amusia-general/.

The Monica study, which showed that tone deafness is not a myth, is "Congenital Amusia: A Disorder of Fine-Grained Pitch Discrimination" by Isabelle Peretz, Julie Ayotte, Robert J. Zatorre, Jacques Mehler, Pierre Ahad, Virginia B. Penhune, and Benoit Jutras (*Neuron*, 2002).

In a large-scale study, Peretz discovered that amusics make up only 2.5 per cent of the population, down from the previously accepted and widely cited 4 percent. But she had yet to publish a paper on her findings when I finished writing this book.

For more on August Knoblauch and the first cognitive model of music, see: "August Knoblauch and Amusia: A Nineteenth-Century Cognitive Model of Music" by Julene K. Johnson and Amy B. Graziano (*Brain and Cognition*, 2003).

The paper that proposed adoption of the MBEA for identifying congenital amusia was "Varieties of Musical Disorders: The Montreal Battery of Evaluation of Amusia" by Isabelle Peretz, Anne Sophie Champod, and Krista Hyde (*Annals of the New York Academy of Sciences*, 2003).

The study that showed that 17 percent of people believe they're tone deaf is "Musical Difficulties Are Rare" by Lola Cuddy, Laura-Lee Balkwill, Isabelle Peretz, and Ronald R. Holden (*Annals of the New York Academy of Sciences*, 2005).

Hutchins first used the slider for "A Frog in Your Throat or in Your Ear? Searching for the Causes of Poor Singing" by

Sean Hutchins and Isabelle Peretz (*Journal of Experimental Psychology*, 2012).

The story of Bach calling Friederica Henrietta an amusa is from *The Cello Suites: J. S. Bach, Pablo Casals, and the Search for a Baroque Masterpiece* by Eric Siblin (House of Anansi, 2009), p. 99.

Track Six
Though George Jones made "A Good Year for the Roses" famous, and Elvis Costello did a fine job with it, Jerry Chestnut wrote it.

My friend's story about our canoe trip is "On Quebec Fur Traders' Trail, with a Paddle" by Alex Hutchinson (*The New York Times*, July 1, 2001).

Track Seven
The opening of this chapter originally appeared, in a slightly different form, as part of "No City for Middle-Aged Men" on the *Toronto Standard* site (August 17, 2011).

The Ulysses S. Grant quip is perhaps apocryphal, as variations of it have been attributed to everyone from Abraham Lincoln to Richie Havens. The *Quote Investigator* website has more: http://quoteinvestigator.com/2013/12/26/two-tunes/.

James Amos quote is from *The Seven Worlds of Theodore Roosevelt* by Edward Wagenknecht (The Lyons Press, 2009). p. 93.

The story about Che dancing the tango to a mambo is from *Che Guevara: A Revolutionary Life* by Jon Lee Anderson (Grove Press, 1997), p. 85.

Full Darwin quote: "From associating with these men and hearing them play, I acquired a strong taste for music, and used very often to time my walks so as to hear on week days the anthem in King's College Chapel. This gave me intense pleasure, so that my backbone would sometimes shiver. I am sure that there was no affectation or mere imitation in this taste, for I used generally to go by myself to King's College, and I sometimes hired the chorister boys to sing in my rooms. Nevertheless I am so utterly destitute of an ear, that I cannot perceive a discord, or keep time and hum a tune correctly; and it is a mystery how I could possibly have derived pleasure from music." *The Autobiography of Charles Darwin* by Charles Darwin and Nora Barlow (Collins, 1958), pp. 61–62.

The Freud quote is from the introduction to *The Moses of Michelangelo* (1914).

William and Henry James: *Musicophilia: Tales of Music and The Brain* by Oliver Sacks (Knopf, 2007), p. 319.

The Nabokov quote is from *Speak, Memory: An Autobiography Revisited* by Vladimir Nabokov (G. Putnam's Sons, rev. ed., 1966), pp. 35–36. In *Musicophilia* (p. 109), Sacks wonders if he might have been joking. My suggestion that, more likely, his tone deafness was a deep regret is based on what author and journalist John Colapinto, who did his master's thesis on Nabokov, told me.

"A Chapter on Ears" was included in *Essays of Elia* by Charles Lamb (1823).

For more on consonance and dissonance, see "The Basis of Musical Consonance as Revealed by Congenital Amusia" by Marion Cousineau, Josh H. McDermott, and Isabelle Peretz (*Proceedings of the National Academy of Sciences*, 2012).

Track Eight

Loui's study on the arcuate fasciculus is "Tone-Deafness: A Disconnection Syndrome?" by Psyche Loui, David Alsop, and Gottfried Schlaug (*Journal of Neuroscience*, 2009).

Diana Deutsch's quote about absolute pitch is from her website: http://deutsch.ucsd.edu/psychology/pages.php?i=215.

"HOMR" is Season 12, Episode 9, of *The Simpsons*.

The section on fMRIs is based on a number of sources, including "Brain imaging: fMRI 2.0" by Kerri Smith (*Nature*, 2012); "What Is fMRI?" (UC San Diego School of Medicine); and "Deceiving the Law" (an editorial in *Nature*, 2008).

The Schulman quote is from "Dangerous Assumptions in Neuroscience," a post on the Oxford University Press blog (http://blog.oup.com), May 2013.

Side Three
Track Nine

For more on music absorption, see: "Absorption in Music: Development of a Scale to Identify Individuals with Strong Emotional Responses to Music" by Gillian M. Sandstrom and Frank A. Russo (*Psychology of Music*, 2013).

Marcus on why music is so engaging: *Guitar Zero*, p. 143.

For more on orchestra conductors, see "Which Part of the Conductor's Body Conveys Most Expressive Information? A Spatial Occlusion Approach" by Clemens Wollner (*Musicae Scientiae*, 2008).

"Facial Expressions of Singers Influence Perceived Pitch Relations" by William Forde Thompson, Frank A. Russo, and Steven R. Livingstone (*Psychonomic Bulletin & Review*, 2010).

Byrne on "lurching and shaking": *How Music Works*, p. 87.
Byrne on low frequencies: *How Music Works*, p. 113.

Evelyn Glennie's "Hearing Essay" is available on her website: www.evelyn.co.uk.

More on the Emoti-chair here: "Designing the Model Human Cochlea: An Ambient Crossmodal Audio-Tactile Display" by Maria Karam, Frank Russo, and Deborah I. Fels (*IEEE Transactions on Haptics*, 2009).

The Randy Bachman quote about B. B. King was an emailed statement that appeared in "Randy Bachman Remembers Meeting B. B. King as Other Canadian Tributes Flow in" by Nick Patch (The Canadian Press, May 15, 2015).

"Facial Expressions and Emotional Singing: A Study of Perception and Production with Motion Capture and Electromyography" by Steven R. Livingstone, William Forde Thompson, and Frank A. Russo (*Music Perception*, 2009).

The section on mirror neurons is largely based on *Mirroring People: The New Science of How We Connect with Others* by Marco Iacoboni (Farrar, Straus and Giroux, 2008); "Eight Problems for the Mirror Neuron Theory of Action Understanding in Monkeys and Humans" by Gregory Hickok (*Journal of Cognitive Neuroscience*, 2009); and interviews with Frank Russo.

In *How Music Works* (p. 200), while talking about songwriting, Byrne — who, needless to say, is not a scientist — says that getting the words right requires finding ones that not only sound right, but feel right on the tongue. "If they feel right physiologically, if the tongue of the singer and the mirror neurons of the listener resonate with the delicious appropriateness of the words coming out, then that will inevitably trump literal sense, although literal sense doesn't hurt," he writes. "If recent neurological hypotheses regarding mirror neurons are correct, then one could say that we empathetically 'sing' — with both our minds and the neurons that trigger our vocal and diaphragm muscles — when we hear and see someone else singing. In this sense, watching a performance and listening to music is always a participatory activity." That's a neat idea, especially given his laments about the professionalization of music elsewhere in the book, if mirror neurons are a real thing in humans.

BAT: "The Beat Alignment Test (BAT): Surveying Beat Processing Abilities in the General Population" by John R. Iversen and Aniruddh D. Patel (*Proceedings of the 10th International Conference on Music Perception & Cognition,* 2008).

Track Ten
This opening of this chapter originally appeared, in a slightly different form, in *Maisonneuve* (Winter 2013) as "Klondike Creative Class" by Tim Falconer.

Something about that song from *Hair* must have resonated with the pre-pubescent mind because many years later, I discovered my experience was not unique as I had imagined. I met a woman whose much older sister had the album back then and played it a lot so one day during a visit from her grandmother, my friend offhandedly began singing the chorus of "Sodomy." Her mother barked to her brother: "Harry, take Nora to the dictionary." Her father, who wasn't one to listen to lyrics, threw the record in the garbage. When I asked my new acquaintance if she found the dictionary definition of those words scary back then, she replied in a Facebook message, "Completely! I was nine, maybe, reading about sodomy and fellatio. I was still at the age where I thought the guys modeling the underwear in the Sears catalogues were disgusting!"

Further to my point about the words and music matching emotionally: When it came time to work on the fourth Talking Heads album, *Remain in Light*, Byrne believed the melodies and the lyrics had to respond to the band's new musical qualities. "One might even say that the recording process, because it privileged trance-like and transcendent music, was about to affect the words that I would gravitate toward," he writes. "The gently ecstatic nature of the tracks meant that angsty personal lyrics like the ones I'd previously written might not be the best match, so I had to find some new lyrical approach. I filled page upon page with phrases that matched the melodic lines of the verses and choruses, hoping that some of them might complement the feelings the music generated" (p. 160).

I called my Dawson City radio show *Face the Music* because that was the title of the article that was the start of this book ("Face the Music" by Tim Falconer, *Maisonneuve*, Spring 2012).

The origin of "Writing about music is like dancing about architecture" is a matter of considerable debate. I've attributed it to Martin Mull because of the convincing case that the Quote Investigator website has made. Also, Elvis Costello, one of several people who often gets credit for the line, thinks it was Mull. Read more here: http://quoteinves-tigator.com/2010/11/08/writing-about-music/.

"Courtney Barnett Prepares Her Debut Album," by Jon Pareles (The *New York Times*, March 13, 2015).

Marcus on why vocal music is more popular: *Guitar Zero*, p. 137.

Track Eleven
Much of the information about Muzak and George Owen Squier came from *Elevator Music: A Surreal History of Muzak, Easy-Listening, and Other Moodsong* by Joseph Lanza (St. Martin's Press, 1994).

The Mozart effect: "Music and Spatial Task Performance" by J. C. McLachlan (*Nature*, 1993); "The Mozart Effect" by J. S. Jenkins (*Journal of the Royal Society of Medicine*, 2001).

For more on music and stress, see "Music Hath Charms: The Effects of Valence and Arousal on Recovery Following an Acute Stressor" by Gillian M. Sandstrom and Frank A. Russo (*Music and Medicine*, 2010).

"'Happy Birthday' Copyright Invalidated by Judge" by Ben Sisario (*The New York Times*, September 22, 2015).

More on kinesthetic learning here: "Attempted Validation of the Scores of the VARK: Learning Styles Inventory with Multitrait–Multimethod Confirmatory Factor Analysis

Models" by Walter L. Leite, Marilla Svinicki, and Yuying Shi (*Educational and Psychological Measurement*, 2010).

Track Twelve
"Bob Dylan, Extending the Line" by David Remnick (NewYorker.com, February 9, 2015).

You can hear a version of R.E.M's "Losing My Religion" that has been digitally altered to be in major key here: https://youtu.be/y6KmiIq2-m8.

American Bandstand's Rate-a-Record segment: *Rock 'n' Roll Dances of the 1950s* by Lisa Jo Sagolla (Greenwood, 2011), p. 57.

Marcus on liking for familiarity: *Guitar Zero*, p. 126.

I have no doubt that "liking for familiarity" is real and I am sure programmers at Top 40 radio stations have long depended on it for the format's existence. But there must be a limit to it because overexposure can make us turn on a song. With some Top 40 pap, this can happen quickly, but I find it happens even with something as brilliant as Leonard Cohen's "Hallelujah." So many artists have covered this song and it's so overplayed that the only version I can stand to listen to now is Cohen's (and maybe k. d. lang's). Sadly, that includes Jeff Buckley's much-loved, but tragically overplayed, rendition.

"Take Me to the River" was written by Al Green and Mabon "Teenie" Hodges. And as good as Green's version is, the Talking Heads version is the best.

Six and a half decades since the birth of rock 'n' roll: I'm going to go with "Rocket 88," a 1951 recording by Jackie Brenston and His Delta Cats—who were, in fact, Ike Turner and his Kings of Rhythm—as the first rock 'n' roll song, though admittedly not everyone agrees with that.

The show I helped make for CBC Radio's *Ideas* was called "The Ballad of Tin Ears" and originally aired May 6, 2014.

That successful bad singers are almost always men—Patti Smith is perhaps the exception that proves the rule—reveals a lot about the music industry: Women who aren't good singers rarely score recording contracts.

Side Four
Track Thirteen
More on Micah Barnes's Singers Playground here: www. singersplayground.com.

More on Steven Lecky's Vox Method here: www.voxmethod. com.

More on Larra Debly's Sweet Music Lessons here: http:// sweetmusiclessons.com.

Track Fourteen

"Before I Grow Too Old" was written by Dave Bartholomew, Antoine Domino (aka Fats Domino), and Robert Guidry (better known as Bobby Charles, who also wrote "See You Later, Alligator").

According to some song lyric sites, Strummer says, "Kelly, that's a take" not "Okay, that's a take" after he finishes "Silver and Gold." Despite many, many listens to the song, I can't tell for sure what he says, but I can see no one named Kelly thanked in the album's liner notes.

Mithen's attempt to learn to sing: "Singing in the Brain" by Steven Mithen (*New Scientist*, 2008).

The study that showed that "in amusic families, 39% of first-degree relatives have the same cognitive disorder" is "The Genetics of Congenital Amusia (Tone Deafness): A Family-Aggregation Study" by Isabelle Peretz, Stéphanie Cummings, and Marie-Pierre Dubé (*The American Journal of Human Genetics*, 2007).

The paper on the first person identified with beat deafness is "Born to Dance but Beat Deaf: A New Form of Congenital Amusia" by Jessica Phillips-Silver, Petri Toiviainen, Nathalie Gosselin, Olivier Piché, Sylvie Nozaradan, Caroline Palmer, and Isabelle Peretz (*Neuropsychologia*, 2011).

Track Fifteen

More on the SMART Lab–Royal Conservatory of Music Parkinson's Choir here: http://smartlaboratory.org/ our-choirs/parkinsons-choir.

No-audition choirs: London City Voices www.london-cityvoices.co.uk; A Joyful Noise: www.ajoyfulnoisechoir.ca; Victoria Good News Choir: http://victoriagoodnew-schoir.com.

The study that showed amusics respond to tempo and timbre is "Sensitivity to Musical Emotions in Congenital Amusia" by Nathalie Gosselin, Sébastien Paquette, and Isabelle Peretz (*Cortex*, 2015).

The study on the universal aspects of music is "Music Induces Universal Emotion-Related Psychophysiological Responses: Comparing Canadian Listeners to Congolese Pygmies" by Hauke Egermann, Nathalie Fernando, Lorraine Chuen, and Stephen McAdams (*Frontiers in Psychology*, 2015).

Paul Swoger-Ruston's take on why timbre is undervalued is similar to McAdams's, but slightly different: "Scientists are mostly studying things they can measure; theorists are saying there are lots of other things going on but it's hard to measure; and musicians doesn't care about either—they just want to make music." And, he might have added, a little money.

Track Sixteen

The Steve Martin banjo bit is available here (audio only): https://youtu.be/8Y30YkQMfvU.

"You Can't Play Sad Music on a Banjo: Acoustic Factors in the Judgment of Instrument Capacity to Convey Sadness" by David Huron, Neesha Anderson, and Daniel Shanahan (*Empirical Musicology Review*, 2014).

Byrne on Gothic cathedrals: *How Music Works*, pp. 16–17.
Byrne on CBGB: *How Music Works*, p. 14.

The musical subcultures quote is from *Let's Talk about Love: A Journey to the End of Taste* by Carl Wilson (Continuum, 2007), p. 17.

A study called "Name That Tune: Identifying Popular Recordings from Brief Excerpts" by E. Glenn Schellenberg, Paul Iverson, and Margaret C. McKinnon (*Psychonomic Bulletin & Review*, 1999) tested how quickly listeners recognized short snippets of popular songs. It found that "Performance was well above chance levels for 200-msec excerpts and poorer but still better than chance for 100-msec excerpts" and "it appears that timbre is more important than absolute pitch for identifying popular recordings from very brief excerpts."

Bonus Track

The information on Florence Foster Jenkins, including the Robert Bager quote, is based on *Theatrum Anatomicum (and*

Other Performance Lectures) by Pablo Helguera (Jorge Pinto Books, 2009), p. 42. *Florence Foster Jenkins*, a biopic directed by Stephen Frears starring Meryl Streep and Hugh Grant, is scheduled for release late summer 2016.

For more on Robert Johnson, see "The Importance of Myth in the Story of Robert Johnson" (www.robertjohnsonblues-foundation.org).

Acknowledgements

The idea for this book struck me in the summer of 2007 and a lot of people have helped me with it since then. First of all, thanks to the faculty and my fellow writers—aka The Friends of the Dave—at the Literary Journalism program at the Banff Centre, who didn't laugh when I originally had the idea. Alex Schultz, David Johnston, and Jackie Kaiser gave me expert advice on developing a decent proposal. David Hayes wisely suggested Micah Barnes as the perfect singing coach.

I became convinced that this could be a book after I wrote a feature called "Face the Music," edited by the masterful Drew Nelles, for *Maisonneuve*. Other parts of the book appeared, in a slightly different form, in another piece for that magazine as well as in *Reader's Digest*, on Toronto Standard, and on CBC Radio's *Ideas*, so thanks also to Haley Cullingham, Carmine Starnino, Ivor Tossell, and Sara Wolch.

"Face the Music" was possible only with funding through a Canadian Institutes of Health Research Journalism Award. And while I may not have written quite as much as I hoped I would at the Berton House Writers Retreat, I am indebted to the Writers' Trust of Canada, the Canada Council, and all of Dawson City for three amazing months in the Yukon.

Wendy Glauser once again provided invaluable research; Meghan Walsh helped with transcribing. Ian Pearson, Chris Goldie, Bill Reynolds, Haley Cullingham, and Matt Braga took on the unenviable task of reading the first draft and made many brilliant suggestions. Tilman Lewis, who copyedited the manuscript, was a great help and, among other things, made me realize how often I actually use the word actually. Paul Washington did the index and tapped his vast musical knowledge to help with a couple of definitions. Rudy Lee fact checked a few chapters and once again showed why fact checkers deserve more respect — and money. (But it goes without saying that all errors in this book are entirely mine.) Marta Iwanek, the Authorzilla-tamer, shot my author's photo, and Kourosh Keshiri shot the cover while I sang along to "Time," from Pink Floyd's *Dark Side of the Moon* in his studio. Kelly Crowe and Paul Hamel graciously and generously hosted the house concert. My friends — and my mom and my sisters — stayed enthusiastic about the idea for this book, which was the best kind of encouragement.

Thanks to everyone at House of Anansi, starting with Sarah MacLachlan, who agreed to read my proposal in the first place. But also to Managing Editor Kelly Joseph and

Publicity Director Laura Meyer. Most of all, special thanks to my fabulous editor, Janie Yoon.

Finally, to Carmen Merrifield, who asked a stranger to dance to "Rock Lobster" all those years ago: I melt with you.

Tim Falconer
Toronto
November 2015

Index

TIM FALCONER is the author of three nonfiction books, including *Drive: A Road Trip through Our Complicated Affair with the Automobile* and *That Good Night: Ethicists, Euthanasia, and End-of-Life Care*. In 2010, he won a Canadian Institutes of Health Research journalism award to write about music and health. The result was a 5,500-word piece on amusia that appeared in the Spring 2012 issue of *Maisonneuve* and won a National Magazine Award. He also helped make a radio documentary on the same subject for CBC Radio's *Ideas*. He teaches magazine journalism at Ryerson University in Toronto and creative nonfiction at the University of King's College in Halifax. He lives and listens to music in Toronto.